符号中国 SIGNS OF CHINA

中国酒

CHINESE WINE

"符号中国"编写组 ◎ 编著

中央民族大学出版社
China Minzu University Press

图书在版编目(CIP)数据

中国酒：汉文、英文 /"符号中国"编写组编著. —北京：
中央民族大学出版社, 2024.9
（符号中国）
ISBN 978-7-5660-2337-7

Ⅰ.①中… Ⅱ.①符… Ⅲ.①酒文化—介绍—中国—汉、英　Ⅳ.①TS971.22

中国国家版本馆CIP数据核字（2024）第016973号

符号中国：中国酒 CHINESE WINE

编　　著	"符号中国"编写组
策划编辑	沙　平
责任编辑	罗丹阳
英文编辑	邱　械
美术编辑	曹　娜　郑亚超　洪　涛
出版发行	中央民族大学出版社
	北京市海淀区中关村南大街27号　邮编：100081
	电话：（010）68472815（发行部）　传真：（010）68933757（发行部）
	（010）68932218（总编室）　　　　（010）68932447（办公室）
经销者	全国各地新华书店
印刷厂	北京兴星伟业印刷有限公司
开　　本	787 mm×1092 mm　1/16　印张：9.75
字　　数	135千字
版　　次	2024年9月第1版　2024年9月第1次印刷
书　　号	ISBN 978-7-5660-2337-7
定　　价	58.00元

版权所有　侵权必究

"符号中国"丛书编委会

唐兰东　巴哈提　杨国华　孟靖朝　赵秀琴

本册编写者

刘　勇

前言 Preface

在中华民族五千多年的历史长河中，酒和酒文化一直占据着重要地位。酒是一种特殊的食品，属于物质生活领域，但酒又融于人们的精神生活。在中国文明史中，酒几乎渗透社会生活中的各个领域，中国人饮酒的意义常常超出饮酒行为本身。

中国古人将酒的作用归纳为三类：酒以治病，酒以养老，酒以成礼。各类盛典、重要活动，通常都设佳宴，用酒招待宾客，以表隆重，以示礼节。饮酒的过程，常常是文化的传播过程，其中不乏思想的交流、智慧的展示、信息的分享、感情的联络。数千年来，中国酒文化被演绎得丰富多彩，醇香绵长。

在世界酿酒史上，中国酒更是独树一帜。酒曲酿酒是中国酿酒的精华，中国人用酒曲造酒要比欧洲人早3000多年，这受益于中国悠久的农业文明史。公元前138年张骞出使西域带回葡萄，引进酿酒艺人，中国开始酿造葡萄酒，

During the more than 5,000 years of Chinese history, wine and wine culture have always occupied an important position. Wine is a special form of food belonging to the material category. Yet wine has also entered people's spiritual life. In the course of Chinese civilization, wine has almost penetrated all spheres of social life. The significance of wine drinking often goes beyond the drinking itself.

Ancient Chinese people summarized the role of wine into three categories: to treat diseases, to maintain good health, and to beef up rituals. During all kinds of ceremonies or important civil events, banquets furbished with good wine were usually hosted to entertain the guests, to mark the significance of the occasion and to show the hospitality of the host. The course of wine drinking always involved a process of cultural dissemination, where views were exchanged, wisdoms displayed, information shared and emotions enhanced. During the past several thousand years, Chinese wine culture has been interpreted as a rich and colorful one, like the lingering aroma of wine.

China has been a unique case in the

葡萄酒传入中国比传到法国尚早七八百年。中国的蒸馏技术自18世纪传入欧洲后，使西方自古以麦芽淀粉糖化谷物、再用酵母菌石糖发酵的传统技术大大提高。中国对世界酒业的另外一个贡献是煮酒以防酸败。中国北宋的《北山酒经》中较详细地记述了煮酒加热技术，而西方各国采用煮酒加热技术的时间比中国晚七百多年。

　　"中国的酒世界"与"世界的中国酒"是"中国名片"的两个面。如果想了解中国，就有必要了解中国酒，了解中国独特的酒文化。

wine history of the world. The essence of wine-making in China is to make wine with yeast. The time when Chinese people in ancient times began to use yeast to make wine was 3,000 years earlier than that of the Europeans. This is certainly related to China's long history of agricultural civilization. In 138 B.C., Zhang Qian was sent on a mission to the Western Regions and brought back grapes and wine artisans. Since then China began to produce grape wine. This was some seven or eight hundred years earlier than that in France. When the distilling technology of China was introduced to Europe in the eighteenth century, it greatly improved the traditional brewery technology in the occident, which involved two steps, saccharifying cereals with malt starch and fermenting with microzyme sugar. Another contribution China made to the world was to cook the wine in order to prevent acerbity or deterioration. The book *North Mountain Wine Scripture* published during the Northern Song Dynasty elaborated the wine cooking and heating technology. This was more than seven hundred years earlier than that in the Occident.

　　"The wine world in China" and "Chinese wine in the world" can be regarded as the two sides of "China's business card". If someone wants to learn about China, it is necessary for him/her to learn about Chinese wines and the unique wine culture in China.

目录 Contents

中国酒的历史
History of Chinese Wine 001

中国酒的起源
Origin of Chinese Wine 002

中国酒的历程
Historic Courses of Chinese Wine 013

中国的酒
Chinese Wine 053

白酒
Baijiu 054

黄酒
Huangjiu 059

药酒
Medicinal Wine 064

中国少数民族的特色酒
Special Types of Wine Made by Chinese Ethnic Groups 068

中国酒文化
Chinese Wine Culture.. 073

古代的酒德和酒礼
Wine Ethics and Etiquettes in Ancient Times.......... 074

酒令
Drinking Games 078

酒与民俗
Wine and Folk Customs 081

酒与唐诗宋词
Wine and the Tang Poems and Song *Ci*-poems.... 093

酒与古典文学四大名著
Wine and the Four Famous Chinese Novels 107

酒与古代书画
Wine and Ancient Calligraphy and Painting 119

酒与酒具
Wine and Wine Utensils 126

中国酒的历史
History of Chinese Wine

中国酿酒、饮酒的历史悠长久远，酒之所兴，可追溯至上古。在历史长河中，中国的酿酒技艺日益精进，与之相随的酒文化也日渐丰富。中国人酿造的酒闪烁着智慧的火花，常在不经意间成为酒中之珍，流芳百世。酒在中国社会的历史演进中颇具影响，无论在国际宴饮、礼仪社交，抑或日常生活中都可见其踪影，是中国历史文化不可或缺的一部分。

China enjoys a long history of wine brewing and wine drinking. The emergence of Chinese wine can be traced back to ancient times. In the course of history, the improvement of wine brewing technology in China has always been accompanied by the enrichment of the wine culture. Shining the sparks of wisdom, Chinese wine has often unintentionally become a cherished variety within the world wine community with enduring fame. Wine occupied an important position in the course of Chinese history. Wine has been indispensable for international banquets, social events and daily activities. Indeed, wine has become an integral part of the Chinese history and the Chinese culture.

> 中国酒的起源

约7000年以前，在西安半坡村先民使用的陶器中，有"酉"形罐出现；到新石器时代晚期，已有大量陶制的酒器，如尊、壶、盅、杯等。

• 刻划鸟纹陶尊（新石器时代）
Pottery Container with Bird Patterns (Neolithic Age, 8500-4500 years ago)

> Origin of Chinese Wine

About 7,000 years ago, pottery jars were used by ancestors of the Banpo Village people near Xi'an City. During the late Neolithic Age Period, many earthen utensils, such as wine or offering containers, pots, wine cups and mugs appeared.

The book *Jin Record-Astronomy Chapter* states, "the three stars to the south of the Xuanyuan Constellation are ags of the Wine Star, who were in charge of food and beverage during banquets." Xuanyuan was the name of an ancient Chinese constellation. In ancient times, the Chinese people believed that wine was created by the Wine Star, and that the three Wine Flag stars were in charge of food and beverage during banquetes.

There were many versions of the origin of Chinese wine. According to

• 陶壶（良渚文化）
Pottery Flagon (Liangzhu Culture, 3300 B.C.-2200 B.C.)

古人认为，酒是天上的"酒星"造的。《晋书·天文志》中说："轩辕右角南三星曰酒旗，酒官之旗也，主宴飨饮食。"轩辕是中国古星宿名，古人认为酒旗三星即掌管宴飨的星宿。

archeological findings, scholars had already traced the wine-making history to 9,000 years ago. As natural wine differs from brewed wine, stories about their origins also differ. Although there have been quite some descriptions of natural wine such as the "ape wine" and

• 甲骨文"酒"字、金文"酒"字
Character "wine" in Oracle bone inscriptions and Bronze inscriptions

关于中国酒的起源有多种说法，目前考古发现的酿造酒痕迹，已经使学者将其追溯到9000年以前。自然酒与酿造酒不同，所以分别有其说法。虽然中国古籍中对远古的"猿酒"与"兽乳酒"等自然酒多有记述，但缺乏翔实的考证，故不具严格的文化意义。

一是"猿酒说"：猿猴将果实丢弃或贮藏在石缝树洞中，其糖分自然发酵成酒浆，从旧石器时期的猿人时代饮至新石器时期的先民阶段，距今10000年左右。关于中国猿猴造酒的例证，最有力的是江苏泗洪县双沟镇的"醉猿"化石。1977年，在双沟镇的中新世的砾岩层

the "beast-milk wine" in ancient Chinese books, there have been few verified evidences. So such statements do not bear any cultural significance in the strict sense of the term.

The "Ape Wine Version" claimed that almost 10,000 years ago, apes discarded or stored fruits in stone crannies or tree holes, where the sugar content was naturally fermented into juice. The juice was drunk by ape-men during the Paleolithic Age and by early men during the Neolithic Age. The most potent evidence of Chinese apes making wine could be found in the Dionysopithecus fossil found in Shuanggou Town, Sihong County, Jiangsu Province. In 1977, Chinese experts and scholars found that,

- 砖雕壁画《猿猴造酒》
  Brick Mural Painting *Apes Making Wine*

中国酒坊（图片提供：全景正片）
A Chinese Brewery

中，发现了保留有第1~3上白齿的左上颌骨的猿化石。学者发现其骨骼上浸透有酒的成分。翌年，中国科学院古脊椎动物与古人类研究所的专家李传夔将此化石命名为"双沟醉猿"。后来人类采集蜂蜜等，再模仿猿酒改善果酒，与酿造酒的历史衔接。

二是"兽乳酒说"：指人类进入农耕社会前的远古时期，渔猎到的畜兽乳汁内的蛋白质和乳糖不期自然发酵，形成原始奶酒，再经人工特意制造成酒，此酒在《礼记小院·礼运篇》中被称为"醴酪"。

三是"空桑秽饭说"：晋代江统在《酒诰》中称："有饭不尽，委余空桑，郁积成味，久蓄气芳，

in the conglomerate terrain of Miocene Epoch in Shuanggou Town, there was an ape fossil of the left maxilla with the first to the third upper molars. Experts found the skeleton of the ape was soaked with wine ingredients. In the following year, Li Chuankui, an expert of the Institute of Vertebrate Paleontology and Paleoanthropology (IVPP) under the Chinese Academy of Sciences (CAS), named this fossil the Dionysopithecus Shuangouensis. Later, people collected honey and fruits to simulate the ape wine, and gradually improved it into fruit wine, bridging the gap between natural wine and brewed wine.

The "Beast-milk Wine Version" claimed that in ancient times before mankind started agriculture and farming

本出于此，不由奇方。"较早提出了谷物自然发酵酿酒说。

关于酿造酒也有不同说法。

一、"近9000年说"：有人认为，河南伊川大莘遗址出土的距今8600—8800年的红陶三足钵，是中国目前发现的最早的实用饮酒器具。2004年，中美学者联合研究发现河南舞阳县贾湖遗址出土的陶器上附着酒石酸。宾夕法尼亚大学酿酒史专家麦克戈温（Patrick E. McGovern）根据C14同位素年代测定，得出其年代在公元前7000—公元前5800年。这证明9000年前贾湖人已经掌握了酒的酿造方法，所用原料包括大米、蜂蜜、葡萄和山楂等。这一发现被列为中国20世纪考古重大发现之一。在此之前，世界公认最早的酒是公元前5400年伊朗人酿制的。

二、"约7000年说"：先有1927年瑞典地质学家安特生（Johan Gunnar Andersson）在河南渑池仰韶村首先发现仰韶文化的农耕技术和酿酒技艺，后有陕西眉县杨家村1983年出土的红陶酒具为证；距今约7000年河南裴李岗文化的新石器遗址、

activities, protein and milk sugar in captured beast and animal milk were irregularly and naturally fermented into primitive milk wine, which was further processed to make wine. This wine was called "*Lilao*" in *The Book of Rites-on Ritual Practices*.

The "Leftovers in Mulberry Forests Version" claimed that the book *Admonition on Wine* by Jiang Tong of the Jin Dynasty stated, "leftovers were dumped in mulberry forests, where they gradually fermented and emitted aroma and fragrance. This was how wine was made." This book was the earliest work that propounded the theory of wine-making through the natural fermentation of cereals.

There also existed several different versions of the origin of brewed wine.

The "Close to 9,000 Years Version" claims that the three legged red pottery jars, unearthed from the Dashen Heritage Site in Yichuan County, Henan Province and made between years 8,600-8,800, were hitherto the earliest practical drinking utensils found in China. In 2004, tartaric acid was discovered on potteries excavated from the Jiahu Site, Wuyang County, Henan Province by Chinese and American scholars jointly.

麦子
谷物是酒形成的原材料。
Wheat
Cereals are the raw materials of Chinese wine.

彩陶三足小口壶（裴李岗文化）
Color Pottery Three-legged Pot (Peiligang Culture, 5600 B.C.-4900 B.C.)

山东莒县大汶口文化遗址出土的酿酒器具与河北武安磁山文化遗址出土的系列陶制酒器也支持此说。

三、"约5000年说"（包括

By testing the C14 isotope age, Patrick E. McGovern, a wine brewing specialist from the University of Pennsylvania, concluded that the tartaric acid should have been made between 7000 B.C.-5800 B.C.. This proved that the Jiahu people had already commanded the method for brewing wine some 9,000 years ago. The raw materials used included rice, honey, grapes, hawthorns, etc. This discovery has been listed as one of the major discoveries in China's archeological history of the twentieth Century. Before this, the earliest wine acknowledged by the world was a kind of brewed wine by the Iranians around 5400 B.C. .

The "About 7,000 Years Version" is supported by Johan Gunnar Andersson, a Swedish geologist, who first found the farming technology and wine brewing technology of the Yangshao Culture in Yangshao Village, Mianchi County, Henan Province in 1927. The finding was supported by later findings of red pottery drinking utensils excavated from Yangjia Village, Meixian County, Shaanxi Province in 1983. This theory has also won the support from the Neolithic Site of Peiligang Culture in Henan Province and wine brewing tools found at the Dawenkou Culture Site in Juxian County,

螺丝形灰陶壶（大汶口文化）
Snail-shaped Gray Pottery Pot (Dawenkou Culture, 4500 B.C.-2500 B.C.)

"4200年说"）中的"杜康说"：源自东汉文字学家许慎的《说文解字·巾部》"酒"条，称"古者少康初作箕帚、秫酒"；"尧帝造酒说"也与此年代基本契合；"仪狄说"，源自公元前239年秦国丞相吕不韦的《吕氏春秋》，称"仪狄作酒"，其后有《战国策·魏策二》和《酒诰》等史书支持此说。

Shandong Province, as well as the series of pottery drinking utensils excavated from the Cishan Culture Site in Wu'an County, Hebei Province nearly 7000 years ago.

The "Du Kang Claim" of the "About 5000 Years Version" (including the "About 4200 Years Version") found its origin in the dictionary by Xu Shen, an etymology scholar of the Eastern Han Dynasty. In the article on "wine", it was stated that in ancient times, Shao Kang first used tools and cereals to make wine. The version of "Emperor Yao making wine" basically coincides with this timing. The version of "Yi Di making wine" was found in the book *Lv's Commentary of History* by Lv Buwei, prime minister of the Qin State, in 239 B.C.. This version also found support in history books such as the *Warring States Strategy*—the Wei Strategy II and *Admonition on Wine*.

杜康造酒、尧帝造酒、仪狄作酒

杜康造酒

关于杜康的身份,有多种说法。一说杜康即少康,是夏王朝的第六位君王,一说杜康是周代人。如果说造酒之祖的话,那么夏朝说比较接近。传说杜康将未吃完的剩饭剩菜放置在桑园的树洞中,经发酵后其散发出芳香气味。这就是酒的酿造方法。因此,人们就把杜康当作酿造酒的祖先,"杜康"也成为酒的代名词。

尧帝造酒

尧为上古五帝之一,传说他是真龙化身,对灵气特别敏感。他受滴水潭灵气所吸引,带领大家在附近安居,并发展农业,使得百姓过上了安定的生活。为感谢上苍,尧帝精选出最好的粮食,并用滴水潭的水浸泡,去除杂质,萃取精华,合酿祈福之水。此水清澈纯净、清香悠长,是最早的酒。

仪狄作酒

仪狄是大禹时期的一位绝色女子,她在做饭的时候闻到谷物的芳香,便将其发酵所成的汁液呈与大禹品尝。大禹喝后觉得甘甜无比,并连饮数杯,逐渐上瘾,导致政事耽搁。后来大禹意识到喝酒误事,便刻意戒酒,远离仪狄。

Three Legendary versions of Wine-making in China

Du Kang's Wine

There were several versions of the identity of Du Kang. One version held that Du Kang was Shao Kang, who was the sixth King of the Xia Dynasty. Another version held that Du Kang lived during the Zhou Dynasty. If he were regarded as the founding father of wine-making in China, the former version was probably closer to the truth. It was said that Du Kang would keep the leftover food in tree hollows in his mulberry grove. In time, the fermented food emitted aroma. This was the original wine brewing method. For this reason, people regarded Du Kang as the founding father of Chinese wine and the name "Du Kang" became synonymous with wine.

Emperor Yao's Wine

Emperor Yao was one of the five emperors in ancient times. As legend had it, Yao was the reincarnation of a living loong and he was very sensitive to nimbus. Attracted by the nimbus of the Water-dropping Pond, he led his people to settle down near the pond and engaged in farming activities and led a peaceful life. In order to express their gratitude to Heaven, Emperor Yao selected the best cereals and soaked them in clear pond water, removed the impurities, and extracted the essence and mixed the water together to make offering water for praying. This clean, pure and aromatic water was the earliest form of Chinese wine.

Lady Yi Di's Wine

Yi Di was a beautiful woman during the reign of Dayu. She was cooking when she smelt the aroma of cereals. So, she fermented the cereals and sent the juice to Emperor Dayu to try. Dayu found the juice very sweet and aromatic. After continuous drinking, he developed an addiction and neglected his duties. Later when he realized that drinking the juice could interfere with his work, he decided to stay away from the juice (wine) and Lady Yi Di.

- 表现古代酿酒场景的壁画
  A Mural Painting Depicting Wine Brewing Scenes in Ancient Times

中国酒历史之最

最早记载酒的文字：商代甲骨文。

最早的麦芽酿成的中国酒精饮料：醴。

最早的葡萄酒记载：西汉司马迁《史记·大宛列传》。

最早的药酒生产工艺记载：西汉马王堆出土的帛书《养生方》。

最早的麦芽制造方法记载著作：北魏贾思勰《齐民要术》。

现存最古老的酒：河南商代后期古墓出土的酒，现存于故宫博物院。

最早提出酿酒始于农耕的著作：汉代刘安《淮南子》。

最早提出酒是天然发酵产物的著作：晋代江统《酒诰》。

现已出土的最早成套酿酒器具：山东大汶口文化时期。

最早反映酿酒全过程的图像：山东诸城凉台出土的《庖厨图》画像石。

已发现的最早的蒸馏器：东汉时期的青铜蒸馏器。

最早的酿酒规章：周代的《礼记·月令》。

古代学术水平最高的黄酒酿造专著：北宋朱肱《北山酒经》。

最早记载加热杀菌技术的著作：北宋《北山酒经》。

古代记载酒名最多的著作：宋代张能臣《酒名记》。

古代最著名的酒百科全书：宋代窦苹《酒谱》。

The Historical "Most" of Chinese Wine

The earliest characters that recorded Chinese wine were inscriptions on oracle bones.

The earliest alcoholic beverage made from malt was "*Li*" (sweet wine).

The earliest written record of grape wine was in the "Dayuan Commentary Section" of the *Historical Records* by Sima Qian in the Western Han Dynasty.

The earliest written record of medicinal wine production processes was contained in the silk book *Health Prescriptions* unearthed from the Tomb of the Western Han Dynasty at Mawangdui.

The earliest written record of malt production was contained in the book *Important Arts for the People's Walfare* by Jia Sixie of the Northern Wei Dynasty.

The oldest wine is the wine unearthed from the ancient tombs of the late Shang Dynasty in Henan Province. Now the wine is kept in the Palace Museum in Beijing.

The earliest book raising the argument that wine-brewing originated from agricultural activities is *Book of Prince of Huainan* by Liu An of the Han Dynasty.

The earliest book raising the argument that wine is produced through natural fermentation is the book *Admonition on Wine* by Jiang Tong of the Jin Dynasty.

The oldest set of wine-making tools unearthed was from the Dawenkou Cultural Period in Shandong Province.

The earliest painting depicting the whole wine-brewing process is the stone carving *The Kitchen* unearthed from Liangtai, Zhucheng City, Shandong Province.

The earliest distillers found are the bronze distillers made during the Eastern Han Dynasty.

The earliest rules and regulations on wine-brewing are found in the Zhou Dynasty's *Book on Rites—Monthly Events*.

The earliest monograph of the highest academic standards on *Huangjiu* brewing in ancient times is the book *North Mountain Wine Scripture* by Zhu Gong, Northern Song Dynasty.

The earliest book that recorded the heating and sterilization technology in ancient times is the book *North Mountain Wine Scripture* in the Northern Song Dynasty.

The book containing the most names of wine in ancient times is called *Record of Wine Names* by Zhang Nengchen of the Song Dynasty.

The most famous encyclopedia on wine in ancient times is the book *Map of Wine* by Du Ping of the Song Dynasty.

• 汉画像石《庖厨图》拓片（图片提供：FOTOE)
Rubbings of stone carving *The Kitchen* of the Han Dynasty

> 中国酒的历程

中国古代文献，包括秦书汉简，几乎"无书不言酒"。祭祀和征伐，是古代中国重要的两件大

> Historic Courses of Chinese Wine

Wine was mentioned in almost all ancient Chinese documentation including bamboo books of the Qin and Han dynasties. While worshipping rituals and crusading ceremonies were the two important events in ancient China, feasts with good wine must be present at rituals, ceremonies, conquering crusades and victory celebrations. *Book on Rites* is a collection of views by Confucius scholars

- 孔子像 马远（宋）

孔子：名丘，字仲尼，春秋晚期鲁国人。中国古代著名的思想家、教育家、政治家，儒家学派的创始人。

Portrait of Confucius, by Ma Yuan (Song Dynasty, 960-1279)

Confucius: given name Qiu, courtesy name Zhongni, was a citizen of the State Lu during the late Spring and Autumn Period. He was a famous thinker, educator, politician in ancient China and the founder of Confucianism.

事，凡祭祀、庆典、出征、凯旋等，必设佳宴，必置美酒，酒为礼而设。《礼记》是先秦儒家记录各种礼仪的汇集，《礼记·乡饮酒义》中对酒食的摆放，酒宴中的入座、举杯、举爵、敬祖、答礼等都有一定之规。

孔子的思想博大精深，以礼论酒是孔子的一个重要理念。明代文学家袁宏道称孔子为酒之饮宗，是

on etiquette and manners during the pre-Qin Period. In the chapter on drinking, there is a detailed elaboration on rules such as how the wine should be placed, how to get seated, how to toast others and respond to toasts, how to offer wine to ancestors, etc.

Confucius theory is profound. To discuss wine from the ethical perspective is one of the important concepts of the Confucius' theory. Yuan Hongdao, a famous writer of the Ming Dynasty, regarded Confucius as the drinking master of Chinese wine simply because both wine-drinking and worshipping rituals are closely related to Confucius' theory on ritual. *The Book of Odes (Shijing)* compiled by Confucius contains about three hundred short poems and articles. Among them thirty mention wine. Another book *Commentaries on the Ten States (Shiguo Chunqiu)* carries the following statement, "King Wen and Confucius drank many cups of wine." Confucius emphasized that a state should be governed and managed by etiquette and music. However, "without Chinese wine, any ceremony is not a perfect one." Without wine, etiquette will lose its form of existence. So, Confucius advocated that "wine-drinking is an academic issue

- **青铜觚（商）**
  觚是古代盛酒的一种器具，容量约为2升。
  A Bronze *Gu* (Shang Dynasty, 1600 B.C.-1046 B.C.)

  A *Gu* is a wine container in ancient times, with a capacity of two liters.

- 酒

  酒在古代中国的政治生活中占有颇为重要的地位。

  Wine

  Wine occupied an important position in the political life of ancient China.

因为饮酒与祭祀都同孔子的礼制思想有着密切的联系。孔子编的《诗经》里面短短三百来篇诗歌，有酒的就占了30篇。《十国春秋》载："文王饮酒千钟，孔子百觚。"孔子强调礼乐治国，而"百礼之会，非酒不成"，没有酒，礼就失去了存在的形式。他提倡"饮酒者，乃学问之事，非饮食之事也"，认为饮酒是关系以德治国、人民安居乐业的事情。

在当时，因为"酒礼"未到，还会引发生战争。如史载"鲁酒薄而邯郸围"，不仅反映出酒在当时已经深入各国政治生活，更透露出春秋时代就已经讲究酿酒的质量。有一次，楚宣王宴请诸侯，鲁国的鲁恭公不但迟到，而且送来的

rather than a diet issue". He was also of the opinion that wine-drinking is related to strategic issues such as how a state should be governed by ethics so that the people enjoy peace and stability.

Wars were triggered by inadequate "wine rites" in ancient times. For example, the story about "Poor wine offered by the State Lu caused the besiegement of the Handan City" shows that wine had penetrated the political life of all states. It also reveals that people already paid attention to the quality of wine during the Spring and Autumn Period. The story goes as the follows: once upon a time, King Xuan of the State Chu entertained nobilities from other states. Duke Gong of the State Lu arrived late and brought some poor-quality wine. This displeased King Xuan. However,

酒还很淡，惹得楚宣王很不高兴。可鲁恭公却说："我是周公王室之后，给你送酒已经是有失礼节和身份了，你还指责酒薄，不要太过分了。"于是不辞而别。楚宣王便发兵与齐国一同攻打鲁国。在这场战役中，赵国的邯郸城也因酒之名而被围困。

Duke Gong said that "I am a descendant of the Zhou Royal Family. It is beneath my dignity to present you wine as gifts. So don't complain about the wine." With that he left without saying goodbye. King Xuan was irritated and launched a war against the State Lu together with the State Qi. During this battle, Handan, capital city of the State Zhao was besieged as a victim.

古代的酒价

前359—前338年：商鞅变法，税重抑商，酒价十倍于成本。

前221—前206年：《秦律》，禁用余粮酿酒，沽卖取利。

前98年：汉武帝采纳理财家桑弘羊的建议，设立"酒榷"官司，实行酒类专卖制度。

前81年：汉代始元六年，官卖酒，每升四钱，这是酒价的最早记载。

The Price of Wine in Ancient China

359 B.C.-338 B.C.: The Shangyang Reform strived to curb business by levying heavy taxes. In such an event, the price of wine rose ten times higher than that of the cost.

221 B.C.-206 B.C.: *Law of the Qin Dynasty* prohibited people from making wine with surplus cereals and selling the wine to gain profits.

98 B.C.: Emperor Wu of the Han Dynasty adopted the suggestion of a then financial expert Sang Hongyang and established the official post "Jiu Que", and applied a monopoly system in the sales of wine.

81 B.C.: Since the sixth year of the Shiyuan Period of the Han Dynasty, wine became the exclusively dealt commodity by the government at the price of four Qian per liter. This was the earliest record on the price of wine.

自天汉三年（前98年），汉武帝刘彻采纳理财家桑弘羊的建议，设立"酒榷"官司，实行酒类专卖制度。这种制度是保证官府的酒业政策得以顺利实施的必要手段。在国家实行专卖政策、税酒政策或禁酒政策时，都对私酒实行一定程度的处罚。轻者没收酿酒器具、酿酒收入，或罚款，重者处以极刑。

魏武帝曹操喜欢饮酒。他在征战中"煮酒论英雄"，其诗歌《短歌行》成为"以酒论政"的千古绝唱。但是，曹操不喜欢别人饮酒，尤其是在他当政以后。为了江山社稷，他发布了《禁酒书》。孔子的20世孙孔融则爱与曹操作对，认为"酒何负于政哉！"

In the third year of the Tianhan Period (98 B.C.), Emperor Wu of the Han Dynasty Liu Che adopted the suggestion by a then financial expert Sang Hongyang, established the official post of "Wine Monopoly" (*Jiuque*), and applied a monopoly system on wine. This system was a necessary measure for the smooth implementation of government policies on wine. When implementing the monopoly policy, taxation or ban on wine, the government always imposed certain penalties on underground or private wine brewing activities. In case of a minor violation, brewing tools, apparatus or sales incomes would be confiscated, and fines imposed. In case of a severe violation, the incumbent would receive penalties up to the death sentence.

- **曹操像**

曹操，字孟德，东汉末年著名的军事家、政治家、诗人。他是三国时期魏国的奠基人，其子曹丕称帝后，追尊其为"魏武帝"。

A Portrait of Cao Cao

Cao Cao, courtesy name Mengde, was a renowned militarist, politician, as well as a poet during the late years of the Eastern Han Dynasty. He was the founding father of the State Wei during the Three Kingdoms Period. When his son Cao Pi took over the throne, he was posthumously bestowed the title of "Emperor Wu of Wei" by his emperor son.

《短歌行》

对酒当歌,人生几何?

譬如朝露,去日苦多。

慨当以慷,忧思难忘。

何以解忧?唯有杜康。

青青子衿,悠悠我心。

但为君故,沉吟至今。

呦呦鹿鸣,食野之苹。

我有嘉宾,鼓瑟吹笙。

明明如月,何时可掇?

忧从中来,不可断绝。

越陌度阡,枉用相存。

契阔谈䜩,心念旧恩。

月明星稀,乌鹊南飞。

绕树三匝,何枝可依?

山不厌高,海不厌深。

周公吐哺,天下归心。

译文:

面对美酒应该高歌,人生短促日月如梭。

好比晨露转瞬即逝,失去的时日实在太多!

席上歌声激昂慷慨,忧郁长久填满心窝。

靠什么来排解忧闷?唯有狂饮方可解脱。

那穿着青领(周代学士的服装)的学子哟,你们令我朝夕思慕。

正是因为你们的缘故,我一直浅吟低唱着那歌。

阳光下鹿群呦呦欢鸣,悠然自得啃食在绿坡。

一旦四方贤才光临舍下,我将奏瑟吹笙宴请宾客。

当空悬挂的皓月哟,你运转着,永不停止;

我久蓄于怀的忧愤哟,突然喷涌而出汇成长河。

远方宾客踏着田间小路,一个个屈驾前来探望我。
彼此久别重逢谈心宴饮,争着将往日的情谊诉说。
明月升起,星星闪烁,一群寻巢乌鹊向南飞去。
绕树飞了三周却没敛翅,哪里才有它们栖身之所?
高山不辞土石才见巍峨,大海不弃涓流才见壮阔。
只有像周公那样礼待贤才,才能使天下人心都归向我。

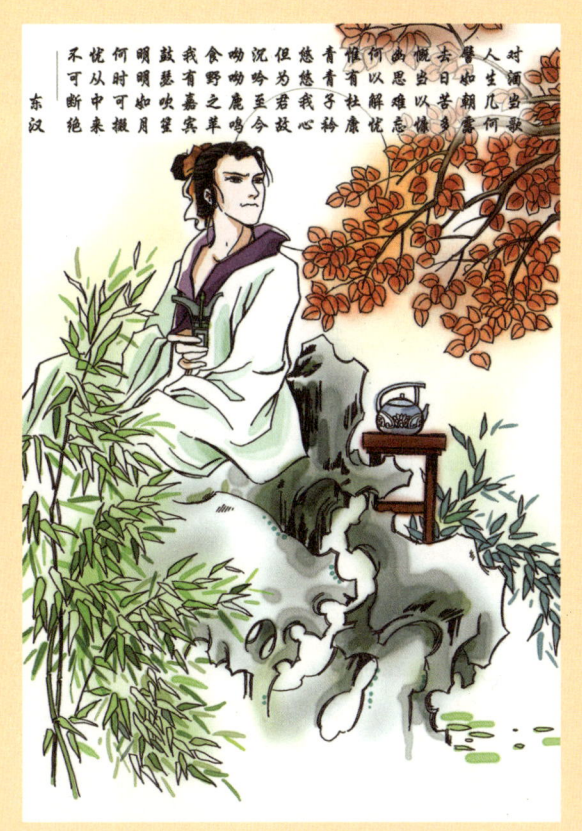

- 短歌行
  The Short Melody

The Short Melody

Wine before us, sing a song.
How long does life last?
It is like the morning dew,
Sad so many days have passed.
Sing hey, sing ho!
Deep within my heart, I pine,
Nothing can dispel my woe,
Save Du Kang, the perfect wine.
Blue, blue the scholar's robe,
Long, long for him I ache.
Preoccupied by you, my lord,
Heavy thoughts for your sake.
For each other cry the deer,
Nibbling grass upon the plain.
When a good friend visits me,
We will play the lyre once again.
In the sky, the moon is bright,
Yet I can reach it never.
In my heart such sorrow dwells;
Remaining with me ever.
In the fields, our paths crossed,
Your visit was so kind.
Together after our long parting,
Your favors come to mind.
Clear the moon, few the stars;
The crows in southward flight.
Circling three times around the tree,
No branch where to alight.
What if the mountain is high,
Or how deep the sea?
When the duke of Zhou greeted a guest,
In his service all wished to be.

在西汉史学家刘向的《战国策》《汉书·食货志》，晋代江统的《酒诰》，以及北魏贾思勰《齐民要术》中，都有大量关于酒生产和酒文化的记载。其中，比较精彩的是王羲之的"曲水流觞"。东晋永和九年（353年）暮春三月三日，大书法家王羲之邀请知名文人学士谢安、孙绰等四十一人，聚会于风景佳丽的兰亭春游宴饮。他们列坐在蜿蜒曲折的溪水边，酒杯顺流漂游，即席赋诗不出者，罚酒三大觥，此谓之"曲水流觞"。事后，王羲之乘兴挥毫，为这些诗文写下了誉满千古的《兰亭集序》。

Emperor Wu of Wei Cao Cao was very fond of wine. Even during battles, he would "award heroes with wine". His poem *The Short Melody (Duange Xing)* has been a poetic masterpiece on "wine politics" through the ages. In spite of his love for wine, Cao Cao disliked others to drink, especially after he took power. For the sake of the administration, Cao Cao issued the "Ban on Wine". Kong Rong, the twentieth generation grandson of Confucius, disagreed with Cao Cao's order. He claimed, "What kind of politics has been delayed by wine?"

Many records on wine production and the wine culture were found in books by ancient scholars such as

- 《兰亭集序》【局部】王羲之（东晋）
Preface for Collections of Poems Composed at the Orchid Pavilion [Part], Wang Xizhi (Eastern Jin Dynasty, 317-420)

• 曲水流觞图（明）
Floating Wine Cups in Winding Stream（Ming Dynasty, 1368-1644）

Liu Xiang, a famous historiographer of the Western Han Dynasty (206 B.C. - 25 A.D.) in his books *Warring States Strategy* and *History of the Han Dynasty—On Economics*, Jiang Tong's book *Admonition on Wine* in the Jin Dynasty, and Jia Sixie of the Northern Wei Dynasty in his book *Qimin Yaoshu: An Agricultural Encyclopedia of the Sixth*

• 漂游在流水中的羽觞
Wine Cups Floating in the Stream

兰亭
The Orchid Pavilion

Century. One excellent piece was the story of drinking wine beside the winding stream. On the third day of the third month of the lunar year of the ninth year of the Yonghe Period during the Eastern Jin Dynasty (353), Wang Xizhi, a famous calligrapher, invited forty-one friends including literati Xie An and Sun Chuo, for an excursion at the Orchid Pavilion where the landscape was beautiful. Sitting beside a winding stream, they started to play the game of drinking from wine cups flowing down the stream. Those who failed to compose a poem impromptu had to drink three cups of wine as a penalty. All the participants had a wonderful time. Later, Wang Xizhi wrote a preface for the collection of poems composed on that day. This famous *Preface for Collections of Poems Composed at the Orchid Pavilion* became a famous literature masterpiece in history.

鸿门宴

鸿门宴是指公元前206年于鸿门（位于陕西西安）举行的一次宴会，参与者是两位反抗秦军的领袖项羽和刘邦。

届时，项羽和刘邦各自攻打秦朝的部队，而兵力较弱的刘邦部队先破秦朝都城咸阳，抢占了称王的先机，令项羽勃然大怒。项羽没有立即攻打刘邦，而是摆下一桌酒席宴请刘邦。这就是历史上闻名的鸿门宴。宴上不乏美酒佳肴，但却暗藏杀机。整个鸿门宴的过程跌宕起伏，险象环生，刘邦屡屡处于危局，却次次都能化险为夷。

先是项羽的亚夫范增"数目项王，举所佩玉玦以示之者三"，暗示项羽杀刘邦，但是"项王默然不应"；再是范增见项羽不为所动，便安排项庄舞剑为酒宴助

● **项羽像**

项羽名籍，字羽，秦末起义军领袖。秦亡后自立为西楚霸王，统治黄河及长江下游的梁、楚九郡，后在楚汉之争中为汉王刘邦所败，在乌江自刎而死。与他有关的历史典故《霸王别姬》是京剧中的经典曲目。

A Portrait of Xiang Yu

Xiang Yu, given name Ji, courtesy name Yu, was a rebel leader during the late Qin Dynasty. After the demise of the Qin Dynasty, Xiang Yu claimed to be the King of the Western Chu State, ruling over nice counties along the Yellow River and lower reaches of the Yangzi River areas. Xiang Yu was later defeated by Liu Bang of the Han State. Xiang Yu committed suicide at the Wujiang River. A famous historical allusion related to Xiang Yu *Farewell My Concubin*e, had been staged as a classical piece of Peking Opera.

兴，趁机杀害刘邦，这就是那句著名的"项庄舞剑，意在沛公"的来历，不料项伯出面与项庄对舞，保护了刘邦；三是在危急关头，刘邦部下樊哙闯入，指责项羽"未有以应"，项羽无言以对，刘邦乘机一走了之，躲过了这场劫难。

鸿门宴对秦末农民战争及楚汉之争都产生了重要的影响，被认为间接促成了项羽溃败，使刘邦成功建立汉朝。后人常用"鸿门宴"来比喻不怀好意的宴会。

Hongmen Banquet

Hongmen banquet was a banquet given at Hongmen (current day Xi'an City, Shaanxi Province) in 206 B.C. The attendees included Xiang Yu and Liu Bang, leaders of the two armies fighting jointly against the Qin-dynasty army.

At the time, Xiang Yu and Liu Bang were partners in leading their respective troops to fight the Qin Army. Liu Bang whose troops were weaker defeated the Qin troops first and took over Xianyang, the capital city of the Qin Dynasty, and seized the opportunity to claim the rule. Xiang

● 刘邦像

刘邦，出身平民，秦朝时起兵于沛（今江苏沛县），称沛公。秦亡后被封为汉王。后于楚汉之争中打败项羽，成为汉朝开国皇帝，史称"汉高祖"。

A Portrait of Liu Bang

Liu Bang , born in a rank-and-file family, led the rebellion in Pei (present-day Peixian County, Jiangsu Province), and claimed to be Peigong (Lord Pei). After the demised of the Qin Dynasty, Liu Bang was bestowed the King of the Han State. Later in the war against the Chu State, Liu Bang defeated Xiang Yu and became the first emperor of the Han Dynasty. He was called by later generations as the "Senior Ancestor of the Han Dynasty".

Yu got furious. However, instead of attacking Liu Bang, Xiang Yu threw a banquet to entertain Liu Bang. This was the famous Hongmen Banquet in history. Although there were plenty of delicious food and wine at the banquet, there were also many hidden crisis. The banquet proceeded with many critical moments with Liu Bang constantly exposed to life-threatening dangers. Somehow Liu Bang managed to escape from the dangers.

- 壁画《鸿门宴图》（汉）
Mural Painting *Hongmen Banquet* (Han Dynasty, 206 B.C.-220 A.D.)

The first critical moment came when Fan Zeng, the senior adviser of Xiang Yu, hinted to Xiang Yu by lifting his jade pendant to give the order to kill Liu Bang. But Xiang Yu remained silent and didn't issue the order. Then Fan Zeng arranged for Xiang Zhuang (an officer) to perform the sword dance as a cover for his attempt to kill Liu Bang. (This is the origin of the famous Chinese proverb "Xiang Zhuang performs the sword dance, aiming at Lord Pei" or "act with a hidden motive".) To Fan Zeng's surprise, Xiang Bo (a senior officer of Xiang Yu) also rose to perform a duet sword dance with Xiang Zhuang and this protected Liu Bang. The third critical moment came when Fan Kuai (an officer in Liu Bang's camp) barged into the banquet, exposing Xiang Yu's ill-intention with convincing arguments. This made Xiang Yu speechless. Liu Bang took this opportunity to leave the banquet. Another crisis was avoided.

The Hongmen Banquet casted an important influence on farmer's rebellion against the Qin Dynasty and the war between the Chu and Han States. It was deemed as having contributed indirectly to Xiang Yu's eventual failure and the success of Liu Bang in establishing the Han Dynasty. "Hongmen Banquet" has been quoted metaphorically to refer to those ill-intentioned banquets.

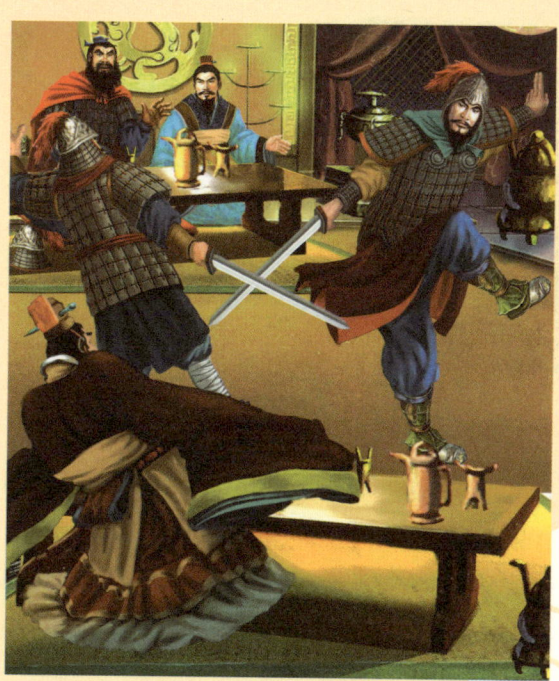

- 项庄舞剑，意在沛公
Xiang Zhuang Performs the Sword Dance, Aiming at Lord Pei

《齐民要术》

《齐民要术》是北魏时期杰出农学家贾思勰所著，是中国第一部农业百科全书，也是现存最完整的中国古农书。本书成书于北魏末年，作者在总结前人经验的基础上，对公元6世纪以前黄河中下游农牧业的生产经验、食品加工与贮藏、野生植物的利用等进行了系统的总结。

Qimin Yaoshu: An Agricultural Encyclopedia of the Sixth Century

The book *Qimin Yaoshu: An Agricultural Encyclopedia of the Sixth Century* was written by Jia Sixue, an excellent agriculturalist of the Northern Wei Dynasty. This is the first encyclopedia on agriculture in China, and is the most complete ancient book on agriculture that survived. Completed during the late Northern Wei Dynasty, the book presented a systematic review of agricultural production, food processing and storage, and wild plants utility in the middle and lower reaches of the Yellow River prior to the sixth century based on the experiences of predecessors.

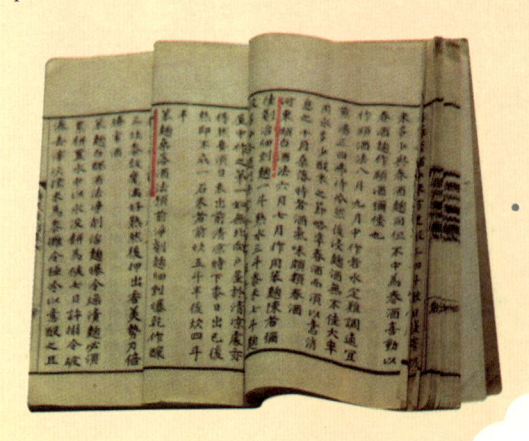

- 《齐民要术》贾思勰（北魏）
书中有大量关于酒生产和酒文化的记载。
Qimin Yaoshu by Jia Sixie (Northern Wei Dynasty, 386-534)
The book contains many records of wine production and wine culture.

晋武帝司马炎好酒，亦"因酒识才"。给他喂马多年的姚馥爱酒，故他一时高兴便封官许愿："任命你为朝歌县令。"可是姚馥却说："我还是给陛下喂马吧！只要陛

Sima Yan (Emperor Wudi of the Jin Dynasty) was fond of wine, and he also "identified talents through wine". Yao Fu who was Sima Yan's horse groom for many years was a wine lover. One day, Sima Yan was very pleased, so he said

下常赐美酒，以乐余年就心满意足了。"司马炎一听便说："喜欢喝酒就好，那就改任酒泉太守吧。"君言如山，司马炎就是如此治国的。东晋大文学家陶渊明写过《饮酒二十首》，他的诗中有酒，酒中有诗，其诗篇与他的饮酒生活皆有名气，为后世歌颂。

to Yao Fu, "I am going to appoint you as the magistrate of Zhaoge County." But Yao Fu responded: "I just want to be your horse groom. I would be very happy if you often give me some good wine." At that, Sima Yan said: "OK, then! Since you like wine, I will appoint you as the mayor of Jiuquan City which is a famous wine production city." This was Sima Yan's approach to governance and he honored his words. Tao Yuanming, a famous literati of the Eastern Jin Dynasty, wrote twenty poems on wine drinking. There was wine in his poems and poetic sentiments in his wine. He earned his fame both for his poems and his drinking style.

- 《渊明醉归图》张鹏（明）

  陶渊明，东晋著名的诗人、文学家、辞赋家、散文家。他一生嗜酒，酒不仅渗入他的生活，也是他文学创作的灵感来源之一。

  Tao Yuanming Returning Home Drunk, by Zhang Peng (Ming Dynasty, 1368-1644)

  Tao Yuanming was a famous poet, literati, verse and prose writer of the Eastern Jin Dynasty. He loved wine all his life. Wine had not only penetrated his life, but also contributed to his literal inspiration and creativity.

• 窖藏黄酒
Huangjiu Stored in Cellar

唐代饮酒之风遍及整个社会，众多酒家（酒肆）遍布城乡各地。唐代酒的种类不少，在唐诗中酒的称谓繁多，有醥（piǎo，清酒）、醪（láo，浊酒）、醴（甜酒）、圣（苦酒）、醍（tí，红酒）、醙（sōu，白酒）、浮蚁（酒沫）、椒浆、腊酒、壶浆、醅（pēi，未过滤的酒）、醁（lù，美酒）等。

唐人常饮的酒大多是大米酿成的黄酒，又分清酒和浊酒。浊酒的酒液中，常有细细的墨绿色米渣浮在上面，故又称"绿蚁"。由于浊酒酿造时间短，制作方便，所以很普及。清酒酿造期长，酒液清澄，味

During the Tang Dynasty, wine drinking was prevalent in the whole society. Wine bars mushroomed both in the urban and rural areas. Just as there were many different types of wine, there were also different names of wine in the Tang poems, such as *Piao* (clear wine), *Lao* (turbid wine), *Li* (sweet wine), *Sheng* (bitter wine), *Ti* (red wine), *Sou* (Baijiu), *Fuyi* (foam), *Jiaojiang, Lajiu, Hujiang, Pei* (unfiltered wine), *Lu* (fine wine), etc.

A common drink for people in the Tang Dynasty was *Huangjiu* which was further differentiated into the clear type and the turbid type. The turbid wine always had a floating layer of deep green residue on the surface, so it was also

道也更醇厚，但产量少，故价格较贵，有"金樽清酒斗十千"之说。

另外，史料还记载，唐太宗破高昌时获得酿酒珍品马乳葡萄及其酿造方法，酿成的酒芳香浓郁，被诗

called the "Green Ant" wine. Due to the short production time required, this turbid wine was very popular. In contrast, the clear wine which took a longer time to brew was only produced in small quantity. Hence the clear wine was more expensive, as described in some poems, "clear wine in a golden cup is worth a fortune".

According to historical documentation, when Emperor Taizong (Li Shimin) of the Tang Dynasty seized the Gaochang City, he obtained some precious raw materials for making wine, the Horse Milk Grapes and the brewing recipe. This wine thus brewed was very aromatic and fragrant. So it was praised by a famous poet Liu Yuxi as being "more aromatic than the legendary wine *Wuyunjiang* (Five-cloud Juice)". There were also some unique types of wine brewed during the Tang Dynasty such as wormwood wine, thatch wine and chrysanthemum wine. These were wine with wormwood, thatch or chrysanthemum. What's more, medicinal wine was also popular during the Tang Dynasty. People soaked medicinal materials in wine, then braised and steamed the wine to make medicinal wine. The most popular medicinal wine at the time was the "Songlao wine". By mixing the turpentine, pine needles and

- 《夜宴图》【局部】佚名（宋）

  此图取材于唐代十八学士夜宴的典故，描绘的是十八个文人雅士于花木葱郁的庭院中秉烛宴饮的场面。雅士们分三席而坐，觥筹交错，各显醉态；仆童侍女忙于搀扶、侍候饮者。

  The Night Feast [Part], by an Anonymous Painter (Song Dynasty, 960-1279)

  The painting depicts the scene of eighteen literati feasting at night in a beautiful garden under candle light. Sitting around three tables, the literati exchanged toasts, played games and were invariably drunk. Young waiters and waitresses were busy helping them and attending to their needs.

人刘禹锡赞为"味敌五云浆"。五云浆是传说中的极致美酒。唐代还有许多独特的酒，如艾酒、屠苏酒、菊花酒等，就是加入艾叶、屠苏、菊花的酒。除此之外，还流行药酒，在酒中加入药材，采用浸泡、蒸煮等方法制作。最为流行的药酒叫"松醪酒"，是将松脂、松叶和松花加入酒中酿制而成，被认为"久服轻身、益气力、延年"。

唐代的酒文化盛行，人们把饮酒赋诗当作最好的消遣娱乐方式。文人有了酒，更显其文采；酒有了文人，更显其价值，可以用"诗酒人生"来概括当时文人的生活。如大诗人白居易所作的两千多首诗里，涉及酒的就有九百多首。

宋代已经出现蒸馏器，但用来蒸酒还缺乏确切的记载。宋代城市里的酒楼也非常多，各家也有招牌酒，如宋词里常常提到的"眉寿""流霞""玉醑"等。在中国古代酿酒史上，学术水平最高、最能完整体现黄酒酿造科技精华，在酿酒实践中最有指导价值的酿酒专著之一便是北宋末期成书的由朱肱著的《北山酒经》。

pine flowers in the wine, people believed that it would "make you lose some weight but feel stronger, conducive to longevity".

Wine culture of the Tang Dynasty was popular among scholars and the literati community, who regarded drinking wine while composing poems as the best form of entertainment. While wine often triggered scholars' literal inspiration, scholars promoted the status and the value of wine. A typical scholar's life at the time could be described as "a life of poetry and wine". For instance, in the two thousand poems of Bai Juyi, a great poet of the Tang Dynasty, more than nine hundred were related to wine.

Although distilling devices already existed during the Song Dynasty, there were no precise records of using them for wine-making during that period of time. At the time, there were many wine bars and restaurants in urban centers, all of which offered their own brand-name wine, such as "*Meishou*", "*Liuxia*" and "*Yuxu*". The book *North Mountain Wine Scripture* by Zhu Gong of the late Song Dynasty was a book on wine with the highest academic value. It contained the most complete description of the essence of wine and instructions on wine brewing operations in ancient China.

- 《清明上河图》【局部】张择端（宋）
  Riverside Scene during the Qingming Festival [Part], by Zhang Zeduan (Song Dynasty, 960-1279)

  图为北宋汴京(今河南开封)的都市生活万象。画中大街小巷，店铺林立，其中就有数家酒楼。
  The painting depicts all aspects of urban life at Bianjing (present-day Kaifeng of Henan Province). There are many stores and shops on the streets, including several restaurants.

杯酒释兵权

公元960年，后周大将赵匡胤发动兵变，众将以黄袍加在赵匡胤身上，拥立他为皇帝，这一事件称为"陈桥兵变"。随后，赵匡胤胁迫后周皇帝禅位，并建立宋朝，定都开封，史称"北宋"。

北宋初期，宋太祖赵匡胤为了防止历史重演，再次出现分裂割据的局面，以高官厚禄为条件，解除将领们的兵权。乾德元年（963年）春，赵匡胤在退朝后留下石守信、高怀德、王审琦、张令铎、赵彦徽、罗彦环等高级将领饮酒。在酒宴上，赵匡胤对部下说："若没有诸位的协助，朕也当不了皇帝。如今朕身为天子，却不如之前做节度使快乐。当了这个皇帝，朕从来没睡过好觉。"石守信等人一听，大惊失色："陛下何出此言，如今天命已定，谁敢有异心？"赵匡胤无奈一笑："谁不想要富贵?有一天,你部下也对你黄袍加身,拥戴你当皇帝。纵使你不想造反,还由得着你们吗?"石守信等人立刻跪下磕头:"臣等愚昧,不知如何才能解除皇上之忧,还请皇上为臣等指示一条生路。"宋太祖说道:"人生苦短,犹如白驹过隙,不如在家颐养天年,君臣之间相安而处,不是很好吗?"众臣们磕头答谢道:"谢皇上隆恩!"

第二天,石守信等便上表声称自己有病,纷纷要求解除兵权,宋太祖欣然同意,让他们罢去禁军职务,到地方任节度使。

这一事件因发生在酒宴之上,故史称"杯酒释兵权"。

Depriving Military Power with a Cup of Wine

In 960, Zhao Kuangyin, a Senior General of the Late Zhou Dynasty, launched a mutiny when many officers forced him to take the throne and became the emperor. This incident was referred to as the "Chenqiao Mutiny". Later, Zhao Kuangyin coerced the emperor of the Late Zhou Dynasty to abdicate the throne and founded the Song Dynasty designating Kaifeng as its capital. The new regime was named "Northern Song".

● 赵匡胤像
A Portrait of Zhao Kuangyin

During the early years of the Northern Song Dynasty, in order to prevent similar mutinies from happening and the split of the country, Emperor Taizu (Zhao Kuangyin) traded the military power of senior generals with high-ranking titles and competitive remuneration packages. In the spring of the first year of the Qiande Period (963 A.D.), Emperor Zhao Kuangyin invited many senior generals to a banquet where delicious food and wine were served. Guests invited to the party included veterans such as Shi Shouxin, Gao Huaide, Wang Shenqi, Zhang Lingduo, Zhao Yanhui and Luo Yanhuan. In the course of the meal, Zhao said to the generals: "Without your support, I could not have achieved what we have now, including this throne. However, since I became the emperor, I have not been as happy as I used to be when I was the local commander. I could not sleep well." Shi Shouxin and others went pale on hearing this remark and ventured to ask: "Now that the new regime has been consolidated, who dares to think otherwise?" The emperor responded with a reluctant smile, saying: "Who could refuse the temptation of wealth?

If someday your subordinates impose an imperial gown on you and support you to rebel against the regime, what choice do you have even if you don't want to wage the rebellion?" Shi Shouxin and others immediately fell on their knees and asked: "How could we alleviate your frustration? Please guide us." At that, the emperor responded: "Time flies and life is short. Nothing is better than enjoying life at home and maintaining a harmonious monarch-courtier relationship." The generals bowed to the emperor in gratitude for the guidance.

On the following day, General Shi Shouxin and others submitted their resignation from military positions on the pretext that they were sick. Emperor Taizu gladly accepted the resignation. He re-appointed them to senior positions in local governments.

Because this event was triggered during a wine party, it was later referred to as "depriving military power through a cup of wine".

- 杯酒释兵权
Depriving Military Power with a Cup of Wine

《北山酒经》是阐述较大规模酿酒作坊酿酒技术的典范，而大文豪苏轼的《酒经》则是描述家庭酿酒的佳作。《酒经》言简意赅，把所学到的酿酒方法用数百字完整地体现出来。北宋时期的窦苹写了一本《酒谱》，该书引用了大量与酒有关的历史资料，从酒的起源、酒之名、酒之事、酒之功，以及温克（指饮酒有节）、乱德（指酗酒无

• 传统酒库
A Traditional Wine Cellar

The book *North Mountain Wine Scripture* systematically expounded on the production process in large-scale wineries. Another book by a famous man of letters Su Shi entitled *The Wine Book* described household brewing operations. *The Wine Book* was a very short article, yet it explained in detail the wine-making process in several hundred words. During the Northern Song Dynasty, Dou Ping wrote a book *The Wine Map*. By citing many historical documents and materials on wine-related topics such as its origin, names, production technology, impacts on the human body, moderate drinking, over-drinking, abstention, myths, foreign wine, aroma, drinking cups and utensils and drinking games. In short, the book presented an all-inclusive description of Chinese wine. Another book *Record of Wine Names*, written during the Southern Song Dynasty, fully recorded the names of over one hundred famous varieties of wine in China. These wines were made by imperial families, senior officials, big restaurants, wineries or ordinary households. Most of the names were very elegant.

Although the Yuan Dynasty only lasted less than two hundred years, it witnessed a flourishing period of the

度)、诫失(戒酒)、神异(与酒有关的一些奇异古怪之事)、异域(外国酒)、性味、饮器和酒令等十几个方面对酒及与酒有关的内容进行了多方位的描述。约成书于南宋的《酒名记》则全面记载了北宋时期全国各地100多种较有名气的酒名。这些酒有的酿自皇亲国戚,有的酿自名臣,有的出自著名的酒店、酒库,也有的出自民间,这些酒名大多用字雅致。

元朝的国史虽不及二百年,却是中国乳酒和葡萄酒业发展的鼎盛时期。元朝统治者十分喜爱马奶酒和葡萄酒。据《元史·卷七十四》记载,元世祖忽必烈至元年间,祭宗庙时,所用的牲畜庶品中,酒采用"湩乳、葡萄酒,以国礼割奠,皆列室用之"。"湩乳"即马奶酒。这无疑提高了马奶酒和葡萄酒的地位。至元二十八年(1291年)五月,元世祖在"宫城中建葡萄酒室",更加促进了葡萄酒业的发展。在元代,葡萄酒常被统治者用于宴请、赏赐王公大臣,还用于赏赐外国和外族使节。同时,由于葡萄种植业和葡萄酒酿造业的大发展,饮用葡萄酒不再是王公贵族的

development of milk wine and grape wine. It was recorded that rulers of the Yuan Dynasty were very fond of koumiss (horse milk wine) and grape wine. According to volume seventy-four of the book *History of the Yuan Dynasty*, during the Zhiyuan Period of the Yuan Dynasty (1264-1294), "koumiss and grape wine were presented to imperial monasteries as offerings to ancestors". This had undoubtedly promoted the profile of koumiss and grape wine. In May of the twenty-eighth year of the Yuan Dynasty (1291), Emperor Shizu (Kublai) "set up a grape wine cellar in the imperial garden", an act that had greatly enhanced the development of the grape wine industry. During the Yuan Dynasty, grape wine was often used to entertain senior officials and foreign guests. It was also used to award top-performing individuals. Thanks to the remarkable development made both in the vineyard and the brewery sectors, grape wine gradually became available to ordinary people. The "Brain Wine" appeared in some Yuan operas and literature such as *The Water Margin* and *The Golden Lotus*. It was recorded that "From winter to early spring, imperial generals and soldiers were often treated with meat and

专利，平民百姓也可饮用葡萄酒。元杂剧以及《水浒》《金瓶梅》中出现过"头脑酒"。史载："自冬至后至春日，殿前将军甲士赐酒肉，名曰头脑酒。"喝这种酒的目的在于御寒。从官府到民间，"头脑酒"是相当风行的。在江南，"吴人谓之遮头酒"。

明朝是酿酒业大发展的新时期，酒的品种、产量都大大超过前世。明朝虽也有过酒禁，但大体上是放任私酿私卖的，政府直接向酿

Brain Wine", in order to resist the cold. The Brain Wine was very popular both among government officials and civilians. In southern China, this wine was called "Head-covering Wine" by people.

Ming Dynasty was a new development phase for the winery industry when the types and quantity of wine increased significantly. Although the Ming government issued regulations on the wine industry, wine was still brewed and sold freely. The government directly levied taxes on wine producers

- 《天工开物》之造酒 宋应星（明）

"凡酿酒，必资曲药成信。无曲即佳米珍黍，空造不成。"（酿酒必须依靠酒曲作为催化剂。没有酒曲，即使是好米好黍也酿不出酒。）

Wine Brewing in *An Ancient Encyclopedia of Science and Technology*, by Song Yingxing (Ming Dynasty,1368-1644)

Wine is produced using distiller's yeast as the catalyst. Without the yeast, even the best rice or millet cannot make wine.

酒户、酒铺征税。由于酿酒的普遍化，不再设专门管酒务的机构，酒税并入商税。据《明史·食货志》载，酒就按"凡商税，三十而取一"的标准征收。这极大地促进了蒸馏酒和绍兴酒的发展。明朝人顾起元所撰写的《客座赘语》中，对明代的数种名酒进行了品评："计生平所尝，若大内之满殿香，大官之内法酒，京师之黄米酒……绍兴之豆酒、苦蒿酒，高邮之五加皮酒，多色味冠绝者。"

明朝皇帝中没有不饮酒的。宫廷酿造或采买的酒以及各地上贡的名酒，构成五花十色的"系列"御酒，还有专门的机构叫"御酒房"。

明朝重臣严嵩垮台后，被抄物资清单上的金酒杯、酒盂、酒缸的重量即有一万七千余两，而其实际价值，又绝对不仅是其重量所能显示的。

明成化斗彩瓷器素以小巧玲珑著称，尤以各式酒杯为最优。史载，至万历时期就已有"成杯一双，值钱十万""成窑酒杯，每对至博银百金"的记录，以至入清以后，几乎历朝均有仿制，品种繁多。

and dealers. Because wine brewing was so popular, the government didn't set up a separate wine administration. Wine taxes were merged with commercial taxes. According to the chapter "Food and Commodity" of the book *History of the Ming Dynasty*, wine taxes were levied at the rate of "one in thirty" which was equal to the rate of commercial taxes. This rate greatly promoted the development of distilled wine and Shaoxing wine which was brewed primarily from glutinous rice. In his book *A visitor's Note*, Gu Qiyuan evaluated a dozen of famous wines in the Ming Dynasty. He said, "In my opinion, the excellent wines include the Full House Aroma found in the Imperial Palace, wine brewed according to the method prescribed by the imperial court found in homes of senior officials, *Huangjiu* in the capital … Bean wine and Bitter Wormwood wine produced in Shaoxing, and Wujiapi produced in Gaoyou City."

All the Ming emperors loved wine. An "imperial wine" series was formed comprising wine brewed or purchased by the imperial court, and those presented by local governments as tribute to the imperial court. A special "Imperial Wine Office" was established to manage these wines.

• 斗彩葡萄纹杯、桑葚纹杯（明）
Contending-color Cups with Grape Patterns and Mulberry Patterns (Ming Dynasty, 1368-1644)

明末清初的战乱，使中国的酿酒业陷于停滞。清初的"移民填川"政策又促进了大规模的南北酿酒技术的交流。清王朝实行了严厉的禁酒制度，但在具体执行时却流于形式。只是到了光绪三十年（1904年），为了维持每况愈下的财政运转，在全国实施征税。

清代的酿酒事宜由光禄寺负责。光禄寺下设良酝署，专司酒醴

After Yan Song, a senior official of the Ming Dynasty was removed from his office, the confiscated asset lists included gold cups and wine utensils and jars weighing more than 17 thousand *Liang* (Chinese unit of weight, 1 *Liang* equals 50 grams). The actual values of these utensils were much higher.

During the Chenghua Period of the Ming Dynasty, exquisite color porcelains were produced, topped by various wine cups. It was recorded that "A pair of wine cups is worthy of 10 thousand *Liang* of silver" and "An exquisite pair of wine cups costs one hundred *Liang* of gold". Entering the Qing Dynasty, there were many imitations of wine cups produced during the preceding regimes.

The Chinese wine industry suffered a period of stagnation during the warring period from the late Ming to early Qing. The "Migration to Sichuan Province" Policy pursued in the early years of the Qing Dynasty boosted large-scale exchanges in wine brewers between the south and the north. Qing Dynasty implemented a strict temperance system on wine, although it looked more like a mere formality. Wine taxes were levied across the country since the thirtieth year

• 斗彩菊纹小杯（明）
Contending-color Small Cup with Chrisanthumum Patterns (Ming Dynasty, 1368-1644)

之事。北京西安门内有酒局房24间，设6名酒匠、2名酒尉，取春秋两季的玉泉水，用糯米1石、淮曲7斤、豆曲8斤、花椒8钱、酵母8两、箬竹叶4两、芝麻4两，即可造玉泉美酒90斤。玉泉酒问世以后，成了皇帝的常用酒。宫中御膳房做菜，也常用玉泉酒调料。玉泉酒还用于祭礼，每年正月祭谷坛、二月祭社稷坛、夏至日祭方泽坛、冬至日祭圜丘坛、岁暮祭太庙，玉泉酒都是作为福酒供祭。

清嘉庆帝御制诗《元旦试笔》的注释表明，清朝元旦开笔仪式始

of the Guangxu Period (1904) in order to sustain the declining financial situation.

During the Qing Dynasty, wine production was under the administration of the Liangniang Division, Guanglusi Bureau. The Central Government set up a twenty-four-room brewery near the Forbidden City in Beijing and staffed it with six wine masters and two managers. The brewery produced a type of wine named Jade Spring. Using spring water collected from the Jade Spring Mountain in western Beijing during the spring and autumn seasons, 1 *Dan* of sticky rice (about 50 kilograms), 3.5 kilograms of Huai distiller's yeast, 4 kilograms of bean yeast, a small amount of pepper, bamboo leaves and sesame seeds, the brewery was able to produce 45 kilograms of delicious wine. Jade Spring wine quickly became the emperor's favorite drink. It was also used as a seasoning material in the imperial kitchen. Jade Spring wine was also used as an offering in commemoration rituals such as the Harvest Altar in the first lunar month, the State Altar in the second lunar month, Temple of Earth in Summer Solstice, Circular Mount Altar in Winter Solstice and the Imperial Ancestral Temple on

于雍正，定制于乾隆，仪式于元旦子时在养心殿东暖阁之明窗举行。金质镶嵌珠宝的"金瓯永固"杯是仪式中必不可少的御用酒杯之一，用以盛屠苏酒。

同治、光绪年间，西苑南海瀛台种植了莲荷上万棵。西太后常令小太监采取莲花蕊，加药制酒，为莲花白酒。同时又有菊花白酒。这是清末宫中因西太后偏爱而身价百倍的两大名酒。

New Year's eve.

According to a footnote to the poem *Trying the Brush on New Year's Day* by Emperor Jiaqing, the launching ceremony for painters to start working on New Year's Day commenced during the Yongzheng Period, and was institutionalized during the Qianlong Period of the Qing Dynasty. The ceremony was held in front of the open window of the East-wing Warm Chamber, Palace of Mental Cultivation in the early morning of New Year's Day. The gold wine cups inlaid with pearls and precious stones were indispensable imperial cups for drinking the herbal medicinal wine during the ceremony.

During the Tongzhi and Guangxu periods of the Qing Dynasty, thousands of lotuses were planted in the South Sea of the Western Garden. Empress Dowager would order junior eunuchs to collect lotus stamens to make Lotus *Baijiu* by adding some herbal medicines. She also ordered Chrysanthemum *Baijiu* be made. Because of her special favor for these two types of *Baijiu*, they became two famous and expensive *Baijiu* in the Imperial Palace during the late Qing Dynasty.

Emperor Qianlong's favorite wine included the Tortoise wine, Pine wine

- 嵌珠宝"金瓯永固"金杯（清）
  Gold Wine Cup for the Emperor with Inlaid Jewlry (Qing Dynasty, 1616-1911)

乾隆帝经常服用龟龄酒、松龄太平春酒、健脾滋肾状元酒。慈禧中年后开始饮如意长生酒。

and the Tonic Champion wine. Since her middle age, Empress Dowager Cixi started to drink a Longevity wine.

乾隆千叟宴

"千叟宴"是清宫中规模最大、与宴者最多的盛大御宴之一，在清代共举办过四次。康熙五十二年（1713年），康熙帝在阳春园宴请全国70岁以上老人2417人。此后雍正、乾隆、嘉庆三朝也举办过类似的千叟宴。

乾隆五十年(1785年)，乾隆帝在乾清宫举行了千叟宴，宴会场面空前盛大。前来赴宴的老人约有3000名，其中既有皇亲国戚、前朝老臣，也有民间奉诏进京的老人。乾隆皇帝还亲自为90岁以上的老人逐一斟酒。

按照清朝惯例，每50年举办一次千叟宴。在这场盛宴上，皇家免费提供御厨精心制作的满汉全席，以及皇家贡品酒水。老人们在宴席上举杯畅饮。这场浩大的千叟宴被当时的文人称作"恩隆礼洽，为万古未有之举"。

Banquet for Thousands of Seniors Hosted by Qianlong

The "banquet for thousands of seniors" was one of the largest scale banquets ever held in the Imperial Palace of the Qing Dynasty. Four such banquets were held throughout the Qing

- 乾隆像
  A Portrait of Emperor Qianlong

Dynasty. In the fifty-second year of the Kangxi Period (1713), Emperor Kangxi entertained 2417 senior people over the age of 70 at the Spring Garden. Since then, similar banquets were held during the Yongzheng, Qianlong and Jiaqing periods respectively.

In the fiftieth year of the Qianlong Period (1785), Emperor Qianlong held a grand banquet in the Palace of Heavenly Purity. It was attended by nearly 3000 old men, including members of the imperial family and their relatives, retired officials and ordinary old folks. Emperor Qianlong personally filled the wine cups for each and every old man over the age of 90. It was said that the old man sitting on the main guest seat was one hundred and forty-one years old.

As a tradition, the "banquet for thousands of seniors" was held every fifty years. On this occasion, sumptuous food and wine were provided free of charge to old men attending the banquet, who drank to their heart's content. This grand banquet was named by the then literati community as "an unprecedented measure for prompting social ethics and harmony".

• 乾清宫内
In the Palace of Heavenly Purity

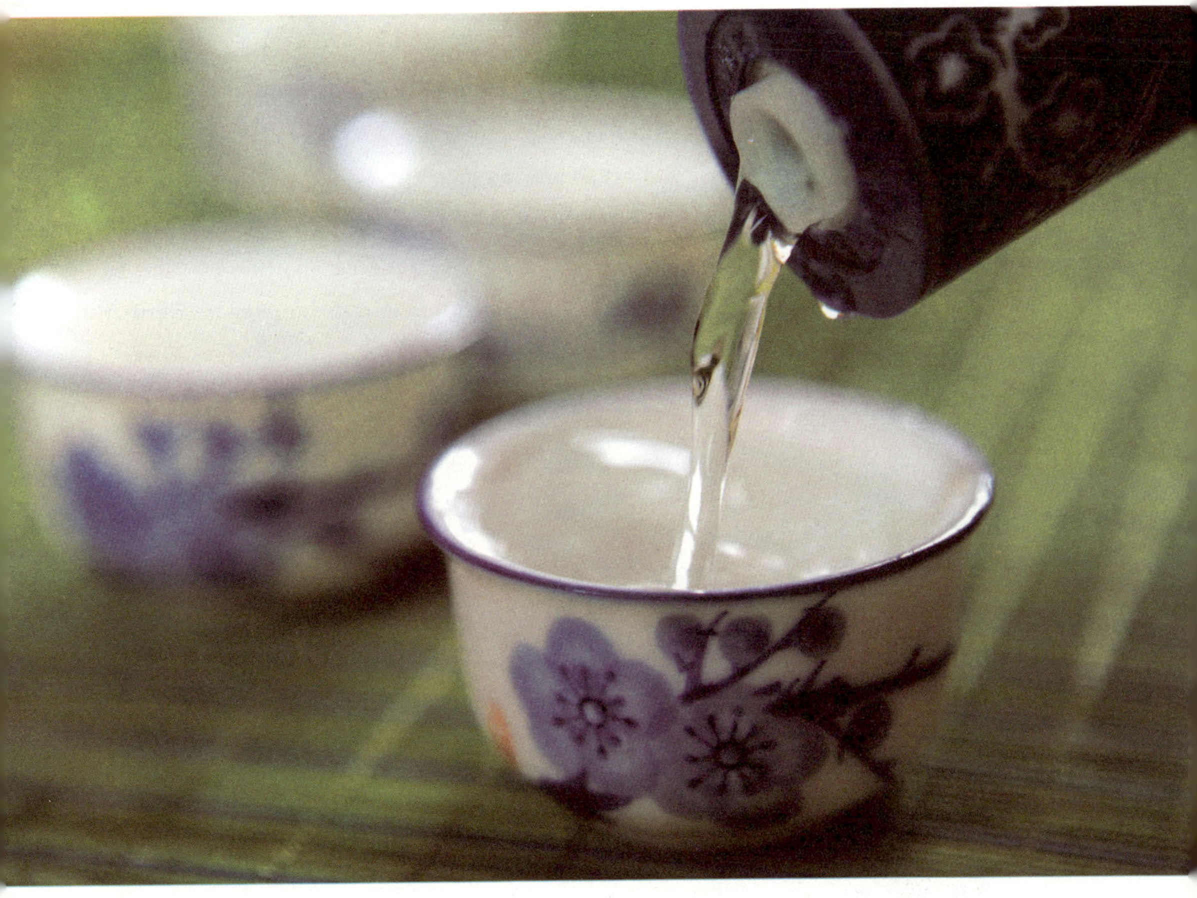

1996年，锦州凌川酿酒总厂的老厂搬迁时，偶然在地下发现了四个木制的酒海（古代存酒的容器），酒海内竟然完好地保存着香气宜人的白酒。这些酒海以红桦构筑，长2.62米、宽1.31米、深1.64米的箱内裱糊了约1500层、内蘸以鹿血的宣纸。这些宣纸上用汉字、满文书写"大清道光乙巳年""同盛金"等字样。专家通过这些记载及其他遗迹和文物，确认这些酒是"同盛金"酒坊在清道光二十五年

When the Jinzhou Lingchuan Wine Company moved to its new premise in 1996, they discovered four wooden wine containers, with well-preserved *Baijiu* in them. These red birch containers were 2.62 meters long, 1.31 meters wide and 1.64 meters deep. The interior was pasted with 1500 layers of classic Chinese paper soaked in deer blood. On the papers were written: "Made in the twenty-fifth year of the Daoguang Period, Qing Dynasty" and "By Tongshengjin Brewery" in Han and Manchurian languages. Further

- 具有一千多年历史的储酒容器——酒海
  A Wine Container with More than One Thousand Years History

（1845年）被封存的，将其命名为"道光廿五"。经英国伦敦吉尼斯总部审核认定，"道光廿五"贡酒是世界目前发现的窖贮时间最长的穴藏贡酒，将其并收入吉尼斯大全。

archeological investigation confirmed that the wine was made and stored by the "Tongshengjin Brewery" in the twenty-fifth year of the Daoguang Period (1845) and was named "Daoguang Twenty-five". Later, the Guinness Headquarters in London confirmed that "Daoguang Twenty-five" was the longest cellar stored wine hitherto discovered in the world. It was recorded in the Guinness World Records.

中国的乳酒

早在新石器时代，原始畜牧业发展起来以后，中国就有了酸奶制成的乳酒。从一些古籍中看，马乳制的酒是乳酒中很重要的一种。汉代朝廷专设有"挏马官"，负责生产马乳酒；《宋史》中也提到高昌（今新疆吐鲁番一带）"马乳酿酒，饮之亦醇"；元代诗人许有壬的《马酒》诗中说："味似融甘露，香疑酿醴泉。新酷撞重白，绝品挹清玄。"

马奶酒是北方少数民族牧民的上等饮料。意大利旅行家马可·波罗曾在其《马可·波罗游记》中说："鞑靼人饮马乳，其色类白葡萄酒，而其味佳，其名曰忽迷思(Koumiss)。"

Chinese Koumiss

As early as during the Neolithic Age, people began to make Koumiss with yogurt with the development of primitive animal husbandry. According to some ancient books, horse koumiss was one of the important koumisses. Han-dynasty government instituted a special post to manage the production of horse koumiss. In the book *History of the Song Dynasty*, it was recorded that "Horse koumiss brewed in Gaochang area (current day Turpan area in Xinjiang) was very aromatic". In his poem *Horse Koumiss*, Xu Youren, a Yuan Dynasty poet wrote: "Horse koumiss tastes as refreshing as dew drops, as aromatic as sweet wine."

Horse koumiss was the high-profile beverage for herdsmen of the northern ethnic groups. The Italian explorer Marco Polo wrote in his book *The Travels of Marco Polo*: "The Tatars drank a type of horse-milk wine whose color is similar to that of the white grape wine and whose texture is so fine. It is called Koumiss."

中国的葡萄酒

1980年在河南省发掘的一个商代后期的古墓中，发现了一个密闭的铜卣。经北京大学化学系相关专家分析，铜卣中的酒为葡萄酒。

根据《史记·大宛列传》记载，汉武帝建元年间，著名探险家张骞从西域带回

- 老葡萄树
Old Grape Vine

来葡萄种子，随后还引来了酿葡萄酒的艺人，中国开始有了葡萄酒。

三国时期葡萄酒的声誉很高。魏文帝曹丕在一封信中说："葡萄酿以为酒……善醉而易醒"。北宋末年朱肱所著的《北山酒经》中所说的葡萄酒，只是以葡萄酒酿造法为工艺，以粮食为主料，以葡萄和杏仁为辅助香料制成的酒。直到元明时期，纯粹的葡萄酒才在内地广泛酿造。除了葡萄酒外，中国古代的果酒还有荔枝酒、椰子酒、石榴酒、枣酒、槟榔酒、甘蔗酒、猕猴桃酒、梨酒等。

20世纪50年代，法国人发现生长在云南的葡萄品种不仅存活，而且得到了优化。从此，跨越了3个世纪的云南品种回到欧洲，重现了历史记忆中的辉煌。这也是中国对全世界葡萄酒业的贡献。

Chinese Grape Wine

In an ancient tomb of the late Shang Dynasty excavated in Henan Province in 1980, a closed bronze jar was found. Analysis conducted by the Chemical Department of Beijing University showed that the wine contained in the bronze jar was grape wine.

As recorded in the "Dayuan Commentary Section" of the Book *Historical Records* during the Jianyuan year of Emperor Wu of the Han Dynasty, the famous Chinese explorer Zhang Qian returned from the Western Regions, and brought back grape seeds as well as wine artisans. Since then China began to produce her own grape wine.

Grape wine enjoyed a prestigious reputation during the Three Kingdoms Period. In a letter written by Emperor Wen of the Wei Kingdom, Cao Pi wrote: "The wine brewed from grapes is easy to get drunk and easy to sober up." The grape wine referred to in the book *North Mountain Wine Scripture* by Zhu Gong of the late Northern Song Dynasty was a wine made through the same brewing process as that of the grape wine. However, the main raw materials used were cereals, with some grapes and almonds as the supplementing material to increase the aroma. It was not until the Yuan and Ming dynasties that pure grape wine was widely produced in China. In addition to grape wine, other types of fruit wine were also produced in China including the wine of lychee, coconut, pomegranate, jujube, areca, sugarcane, kiwi fruit and pear.

In the 1950s, French people found that this grape species not only survived in Yunnan, but was also optimized. From then on, the Yunnan grape was replanted back in Europe after three centuries of time had elapsed, and rescued the glorious memory of history. This was also China's contribution to the global grape wine industry.

中国的啤酒

中国用谷芽酿造醴酒,和巴比伦人用麦芽做啤酒,差不多同时出现于新石器时代,彼此之间是否有联系却无从考察。

中国早在3200年前就有一种用麦芽和谷芽作谷物酿酒的糖化剂,酿成称为"醴"的酒。这种滋味甜淡的酒虽然不叫啤酒,但可以肯定它类似现在的啤酒。《齐民要术》中详细记载了小麦麦芽及饴糖的做法,其麦芽的制造过程与现代啤酒工业的麦芽制造过程基本相同。《隋书·西域国传》还特地记载了党项人把汉人的大麦带回去酿酒的事。只是由于后人偏爱用曲酿的酒,嫌醴味薄,以至于这种酿酒法逐步失传,大约汉代以后,醴被酒曲酿造的黄酒所取代。

Chinese Beer

The time when ancestors of the Chinese people made sweet wine with millet malt and the time when the Babylon people made beer with wheat malt were both around the Neolithic Age. However, no studies have been conducted to show if there were correlations between the two.

As early as 3,200 years ago, ancient Chinese people began to make "sweet wine" using wheat and rice malt with saccharification agents. Although this light and sweet drink was not called beer, it was certainly very similar to beer in modern times. The book *Qimin Yaoshu: An Agricultural Encyclopedia of the Sixth Century* fully expounded the methods for making wheat malt and sugar, which was similar to that of the wheat malt for modern beer. The Western State Section of the book *History of the Sui Dynasty* specially recorded the story of how the Dangxiang people took away barley to make wine. Unfortunately, the method for making this light sweet wine was gradually lost as people seemed to prefer yeast-brewed wine better than malt-made wine. Around the time of the Han Dynasty, "Sweet Wine" was replaced by yeast-brewed *Huangjiu*.

- 啤酒

虽然啤酒在古代中国并未被广泛传播,但是如今俨然成为中国人消耗量最大的酒类之一。

Beer

Although beer was not popularized in ancient times in China, it has certainly become one of the most quantity consumed alcoholic beverages in China today.

民国时期的中国酿酒业有了新的发展。南方盛行黄酒，北方盛行烧酒，西方葡萄酒和啤酒的生产技术传入东部地区。大量名酒则集中在东南沿海地区，而绍兴、汾阳、凤翔、洋河、茅台等特色酒业产区的形成，为中国酒业带来了新活力。

中华民国北京政府执政初期，一方面沿袭清末旧制，保留了清末的一些税种，还制定了一些新的酒政形式，如"公卖制"。民国北

• 绍兴酒馆
浙江绍兴以产黄酒而出名，绍兴黄酒以糯米为原料制成，是中国黄酒的代表。

Wine Restaurants in Shaoxing
Shaoxing City of Zhejiang Province is a famous producer of *Huangjiu*. Using glutenous rice as the raw material, Shaoxing *Huangjiu* is a typical representative of Chinese *Huangjiu*.

During the Minguo Period, new achievements were made in China's wine sector. While *Huangjiu* prevailed in southern China, *Baijiu* prevailed in the north. Meanwhile, wine and beer production technologies were transferred from foreign countries to eastern China. Many famous Chinese wine found their home in coastal cities in China's southeastern regions. The formation of specialized wine production zones such as Shaoxing, Fenyang, Fengxiang Yanghe and Maotai, brought about new vigor and vitality to the Chinese wine sector.

During the early years of the Minguo Period, the Beijing Government implemented a policy on wine taxes retained from the late Qing Dynasty and introduced some new laws to govern the wine sector such as "Public Sales System". The policy started implemented in the fourth year of the Minguo Period (1915). The administrative authorities in charge of this work were the Tobacco and Wine Public Sales Bureaus of the Beijing Government of the Minguo Period and its subsidiaries in the provinces. In June of the sixteenth year of the Minguo Period, the Nanjing Government issued an *Interim Regulation on Public Sales of Tobacco and Wine*, establishing the

京政府实行的酒类"公卖制"始于民国四年（1915年）。负责推行的行政管理机构是民国北京政府的烟酒公卖局和各省的烟酒公卖局。南京国民政府在民国十六年（1927年）六月，公布《烟酒公卖暂行条例》，规定以实行官督商销为宗旨。公卖机关的组织结构与民国北京政府大致相同。公卖费率以定价的20%征收。抗战胜利后，国民政府对某些条例进行了修订，主要目的是提高税率。民国三十五年（1946年）八月，国民政府公布《国产烟酒类税条例》，将国产酒类的税率提高为80%。

中华人民共和国成立后，最先将酒的研究系统提升到文化层面的是于光远先生，他说："广义的文化，包括酒文化的发展，在一定的程度上对我国的经济建设以及人民生活有影响。"

principles of government supervision and merchant sales. The organizational structure of the public sales agency was similar to that of the Beijing Government of the Minguo Period. The Public sales tax rate was twenty percent of the fixed price. After the victory of China's resistance against the Japanese invasion, some provisions were amended in order to raise the tax rate. In August of the thirty-fifth year of the Minguo Period, Nanjing Government promulgated the *Regulations on taxation of Domestic Tobacco and Wine*, which raised the tax rate of domestically produced wine to eighty percent.

Since the founding of People's Republic of China (1949), Yu Guangyuan was the first person to upgrade wine research to the cultural level. Yu said, "To a certain extent, the broad culture, including the wine culture can impact China's economic development and people's livelihood."

中国的酒
Chinese Wine

中国酒历史悠久，工艺独特，在世界诸多酒类中独树一帜。中国酒按制造工艺可分为发酵酒、蒸馏酒、配制酒。黄酒和药酒是中国特有的酒类。

With a long history and unique processing techniques, Chinese wine enjoys an important position in the world wine community. Chinese wine can be divided into fermented wine, distilled wine and mixed wine according to their different production processes. Among them, *Huangjiu* and medicinal wine are special varieties produced exclusively in China.

> 白酒

中国的白酒历史悠久，工艺独特，是世界蒸馏酒中独具一格的一个品类。从古至今，白酒在中国人的生活中占有十分重要的位置，是社交、喜庆、送礼等活动中不可缺少的特殊饮品。

关于白酒的起源目前还没有定论，但在四川彭州、新都先后出土的东汉画像砖上，有生产蒸馏酒作坊的画像，其与四川传统蒸馏酒设备"天锅小甑"极为相似。在唐代文献中，烧酒之名已经出现。李肇于806年写的《国史补》中有："酒则有剑南之烧春。"（唐代称酒为"春"）。田锡的《曲本草》中载："暹罗酒以烧酒复烧二次……蜡封，埋土中二三年绝去烧气，取出用之。"足见唐代已经有了白酒。

> Baijiu

With a long history and a unique processing technique, Chinese *Baijiu* is a unique type of wine in the world wine community. Since ancient times, *Baijiu* occupied a prominent position in Chinese people's daily life, and was regarded as an indispensable special beverage during rituals, ceremonies, festivals and social gatherings.

Although there is not yet a conclusion on the origin of Chinese *Baijiu*, pictures of *Baijiu* distilling mills were found in tombs of the Eastern Han Dynasty in Pengzhou and Xindu of Sichuan Province. This is very similar to the traditional distilling devices used in Sichuan area, metaphorically called the "heavenly wok and rice pot". The term *Baijiu* already appeared in Tang Dynasty documents. The book *Supplementary Records to*

白酒是由高粱、玉米、红薯、麦黍等粮食，以及富含淀粉和糖的其他农副产品和植物为原料发酵、蒸馏而成。酒液无色透明，故称"白酒"。白酒也叫"烧酒"，因为主要是采用烧蒸的工序。白酒的酒精含量很高，一般为30~65度。白酒按糖化、发酵剂、酿造工艺的不同可分为大曲酒、小曲酒、麸曲酒三大类；按酒精含量分类，可分为高度酒和低度酒；按香型分类则有清香、酱香、浓香、米香、兼香等。清香型又称"汾香型"，以

History written by Li Zhao in the year 806 stated, "*Jiannan Baijiu* is a famous Chinese wine." In his book *Materia Medica*, Tian Xi wrote, "Siam liquor is a redistilled Chinese wine, sealed and stored underground for two or three years to get rid of the distilling avor before taken out for serving." These records fully testify to the fact that *Baijiu* already existed in China during the Tang Dynasty.

Baijiu is fermented and distilled from cereals such as sorghum, corn, sweet potato and millet, as well as other agricultural produce and plants with rich starch and sugar contents. Due to the achromatic and transparent color of the juice, the wine is called *Baijiu*, or "distilled *Baijiu*". *Baijiu* is also called distilled *Baijiu* as they are produced through the processes of fermentation and distillation. Chinese *Baijiu* has high alcohol contents, ranging from thirty-sixty percent. *Baijiu* can be divided into three broad categories according to their different saccharification, fermentation and brewing process: *Daqu*, *Xiaoqu* and *Fuqu*. They can also be divided into two categories of high and low alcoholicity. If classified according to their aroma, they can be divided into light-aroma, sauce-aroma, strong-aroma, rice-aroma and mixed-aroma types. Light-aroma *Baijiu*

- 四川彭县（现彭州市）出土的东汉酿酒画像砖

Brick Image of Wine Brewing during the Eastern Han Dynasty unearthed in Pengxian County (Present-day Pengzhou City), Sichuan Province

山西省的汾酒为典型代表；浓香型又称"泸香型"，以四川省的泸州老窖特曲酒为典型代表；酱香型又称"茅香型"，以贵州省仁怀市的茅台酒为代表；米香型以广西桂林三花酒为典型代表。中国酒业泰斗秦含章先生还专门将西凤酒分类为"凤香型"。各种香型各有特色，或香气馥郁，或余香不尽，或醇厚柔绵，或甘润清冽，或回味悠久。

白酒的名称繁多，有的以原料命名，如高粱酒、地瓜酒；有的以产地命名，如茅台酒、兰陵大曲；有的以人名命名，如杜康酒；有的还按发酵、储存时间长短命名，如特曲、陈曲、头曲等。

is also known as *Fenxiang Baijiu*, which is represented by the *Fenjiu* produced in Shanxi Province. Strong-aroma *Baijiu* is also known as *Luxiang Baijiu*, which is represented by *Luzhou Laojiao Tequ* produced in Sichuan Province. Sauce-aroma *Baijiu* is also known as *Maoxiang Baijiu*, which is represented by *Maotai* produced in Renhuai City, Guizhou Province. Rice-aroma *Baijiu* is represented by *Sanhua Baijiu* produced in Guilin City, Guangxi Autonomous Region. Qin Hanzhang, a famous wine expert in China specifically classified the *Xifeng Baijiu* as the *Fengxiang* aroma type. Different aroma types represent different characteristics. Some are strong, others are enduring; some have a very mild texture; others have a sweet, moist and chilly texture; still others have long aftertaste.

Chinese *Baijiu* is named following a variety of principles. Some are named after the raw materials they are made from, such as the Sorghum and the Potato *Baijiu*. Some are named after their production sites, such as the *Maotai* and the *Lanling Baijiu*. Others are named after famous people, such as the *Dukang Baijiu*. Still others are named according to the length of their fermentation and storage, such as the *Tequ Baijiu*, the *Chenqu Baijiu*, the *Touqu Baijiu*, etc.

中国的酒曲酿酒

从有文字记载以来，中国的酒绝大多数是用酒曲酿造的，中国的酒曲法酿酒对于周边国家有较大的影响。中国酿酒的精华是酒曲酿酒。酒曲中所生长的微生物主要是霉菌。对霉菌的利用是中国人的一大发明创造。随着时代的发展，中国古代人民所创造的方法将日益显示其重要的作用。

中国酒曲分为大曲、小曲、红曲、麦曲和麸曲。大曲分为传统大曲、强化大曲（半纯种）和纯种大曲。小曲按接种法分为传统小曲和纯种小曲；按用途分为黄酒小曲、白酒小曲、甜酒药；按原料分为麸皮小曲，米粉曲，液体曲。红曲主要分为乌衣红曲和红曲，其中红曲又分为传统红曲和纯种红曲。麦曲分为传统麦曲（草包曲、砖曲、挂曲、爆曲）和纯种麦曲（通风曲、地面曲、盒子曲）。麸曲分为地面曲、盒子曲、帘子曲、通风、液体曲。

山东诸城凉台出土了一幅汉代的画像石，其庖厨图中的一部分是对酿酒情形的描绘，把当时酿酒的全过程都表现出来了。图中一人跪着正在捣碎曲块，旁边一口陶缸内应为浸泡的曲末，一人正在加柴烧饭，一人正在劈柴，一人在甑旁拨弄着米饭，一人负责将曲汁过滤到米饭中去，并把发酵醪拌匀。有两人负责酒的过滤，还有一人拿着勺子，大概是要把酒液装入酒瓶。下面是发酵用的大酒缸，都安放在酒垆之中。酒的过滤大概是用绢袋，并用手挤干。过滤后的酒放入小口瓶，作进一步陈酿。

- 大曲

大曲的生产工艺流程：小麦→润水→堆积→磨碎→加水拌和→装入曲模→踏曲→入制曲室培养→翻曲→堆曲→出曲→入库贮藏→成品曲。

Daqu

Daqu's production process: wheat→sprinkle water→stacking→grinding→mixing with water→putting in yeast mould→cultivating in yeast chamber→stirring the yeast→stacking the yeast→tredding the yeast→yeast emerging→storage→finished product.

Brewing Chinese *Baijiu* with Yeasts

• 小曲
Xiaoqu

Since the time when written records became available, most Chinese wine has been made with yeasts through fermentation. The Chinese technology of wine-making with yeasts has cast a strong influence on neighboring countries. The essence of Chinese wine-making is yeast-based brewing. Microbes in the yeast are mainly moulds. Using moulds in wine brewing is a Chinese invention. With the advancement of time, technologies invented by the Chinese people in ancient times will increasingly display their significance in modern times.

Yeasts used in Chinese wine can be divided into categories of *Daqu*, *Xiaoqu*, *Hongqu*, *Maiqu* and *Fuqu*. *Daqu* yeasts can be divided into traditional *Daqu*, intensive *Daqu* (semi-pure) and pure daqu. *Xiaoqu* yeasts can be further divided into traditional *Xiaoqu* and pure *Xiaoqu* according to their different ways of seeding. If classified by purpose, the yeasts can be divided into *Xiaoqu* for *Huangjiu*, *Baijiu* and sweet wine respectively. If classified according to the raw materials they use, *Xiaoqu* yeast can be divided into husk, rice flour and liquid types. Red yeast can be divided into *Wuyi* red yeast and ordinary red yeast types, which could be further divided into traditional red yeast and pure red yeast. Wheat yeasts can be divided into traditional wheat yeast (including straw, brick, hanging and explosive types) and pure wheat yeast (including ventilation, floor and box types). The husk yeast can be divided into floor, box, screen, ventilation and liquid types.

On a mural painting unearthed from a Han-dynasty tomb at Liangtai County, Zhucheng City, Shandong Province, there are depictions of the whole process of wine brewing: One kneeling person was mashing the yeast. A pottery jar standing by should be the container for soaking yeast. One person was heating and cooking. Another one was preparing firewood. Still one other person was mixing the rice and one more person was filtering the yeast juice into the rice. Two others were filtering the *Baijiu*. Still one more person was holding a scoop, probably infusing the *Baijiu* into bottles. There were many big jars in the mill. *Baijiu* was probably filtered in silk bags or with hands. The filtered *Baijiu* was put into small bottles for further brewing and storing.

> 黄酒

黄酒作为中国最古老的酒种之一，在酒文化史上占有重要的地位。黄酒因其颜色而得名，又称"老酒"。黄酒与白酒的酿造方法完全不同，它是采用压滤工艺生产的，因而较好地保留了发酵过程中产生的葡萄糖、糊精、甘油、矿物质、醋酸、醛、酯等。它是以谷物（主要是大米、黍米和粟米）为主要原料，经过特定的加工酿制而成的一种低酒精含量的原汁酒，在12°—18°之间。黄酒按原料、酿造方法、风味可分为：江南黄酒（包括韶关黄酒、绍兴加饭酒、状元红、女儿红、花雕酒）、山东黄酒（包括即墨老酒、青酒、兰陵美酒）、福建黄酒（福建老酒和沉缸酒）。

> *Huangjiu*

Huangjiu is one of the oldest Chinese wine varieties, enjoying a prominent position in the history of wine culture. *Huangjiu* is named after its color (*Huang* means yellow and *Jiu* means wine in Chinese). The brewing method of *Huangjiu* is completely different from that of *Baijiu*. *Huangjiu* is made through a pressing and filtering process, which enables the wine juice to retain contents such as dextrose, dextrin, glycerin, minerals, acetic acid, aldehyde and ester generated during the fermentation process. *Huangjiu* is a kind of low alcohol (twelve to eighteen percent) content original-juice wine produced and fermented from cereals (mainly glutinous rice). *Huangjiu* can be divided into many sub-categories according to the different raw materials used, brewing methods

黄酒可提供给人的热量比啤酒和葡萄酒都高得多。黄酒中含有十多种氨基酸，大多数氨基酸是人体不能合成却必需的。据测定，每升黄酒中的赖氨酸的含量，在中外各种营养酒类中最丰富，所以人们把黄酒誉为"液体蛋糕"。由于黄酒酒精含量远远低于白酒等蒸馏酒，且具补血气、助运化、舒筋活血、健脾补胃、祛风抗寒的功能，所以还被用来制药。

优质黄酒的特点：酒色一般为浅黄色，澄清透明有光泽，无悬浮物，不浑浊，无沉淀物，并有明显的黏性。酒香浓郁芬芳，酒味醇厚适口，入口清爽，回味悠长。

employed and different aromas: *Huangjiu* from southern China, *Huangjiu* from Shangdong and *Huangjiu* from Fujian.

Huangjiu can provide people with much more calories than beer and grape wine. *Huangjiu* contains a dozen of amino acids, most of which cannot be generated by the human body but are necessary to the human body. Statistics show, these amino acids contained in *Huangjiu* are more than those in any other tonic wine. So *Huangjiu* has been honored as the "liquid cake". In addition, *Huangjiu* is used for treating illness and making medicine because of its low alcohol content compared to *Baijiu* and other distilled wine and its tonic effects such as enriching blood circulation, helping relax the muscles, strengthening the stomach and the spleen, and treating the cold.

High-quality *Huangjiu* is featured by the following characteristics: light yellow color, clear and transparent luster, no floating substance or sediment and obvious viscosity. Its aroma is strong, thick and enduring, with lasting aftertastes.

- 坛装黄酒
  Huangjiu in a Jar

- 黄酒最适合贮存在陶坛中
  Huangjiu is Best Stored in Pottery Jars

黄酒的饮法

黄酒可带糟饮用,也可仅饮酒汁。后者较为普遍。

传统饮法是温饮,将盛酒器放入热水中烫热,或隔火加温。温饮口味柔和,酒香倍佳。

古人饮酒用注子和注碗,注碗中注入热水,注子中盛酒后,放在注碗中。近代以来,用锡制酒壶盛酒,放在锅内温酒,一般以不烫口(45℃~50℃左右)为宜,但加热时间不宜过久,防止酒精挥发,淡而无味。冬天更盛行温饮。

饮酒时配以不同的菜,则更可领略黄酒的特有风味。以绍兴酒为例:干型的元红酒宜配蔬菜类、海蜇皮等冷盘;半干型的加饭酒宜配肉类、大闸蟹;半甜型的善酿酒宜配鸡鸭类;甜型的香雪酒宜配甜菜类。

温热后杯中加入一两枚话梅,既能改善口感,也可促进血液循环、抵御风寒。

- 黄酒酒具
  Utensils for Drinking *Huangjiu*

How to Drink *Huangjiu*

Huangjiu may be taken with juice and distillers' grains together or the juice alone. The latter is more popular.

Traditionally, *Huangjiu* is served warm, either by putting the wine cups in hot water, or on fire. When heated the texture is milder and more aromatic.

In ancient times, people used an ewer and a bowl to warm up the wine. Wine is contained in the ewer which is put into the bowl containing hot water. In more recent dynasties, a tin flagon replaced the ewer while a cooking wok replaced the bowl. The best temperature of the wine should be between 45 ℃ -50 ℃. It is highly recommended not to overheat the wine, lest the alcoholic content evaporates and the wine will taste plain and less aromatic. This servicing method is more popular in winter.

When drinking the *Huangjiu*, it should be accompanied by various dishes. This will enable the diner to better perceive the special texture of the wine. Taking Shaoxing *Huangjiu* as an example, dry Shaoxing *Huangjiu* should be accompanied by cold dishes such as vegetables and jellyfish. Semi-dry Shaoxing *Huangjiu* should be accompanied by meat and crabs. Semi-sweet Shaoxing *Huangjiu* should be accompanied by chicken and ducks. Sweet Shaoxing *Huangjiu* should be accompanied by sweet dishes.

A couple of plums in the heated wine will improve the taste, accelerate the blood circulation and resist the cold.

> 药酒

　　《汉书·食货志》中记载，"酒，百药之长"。酒与药有密不可分的关系。在远古时代，酒就是一种药，古人说"酒以治疾"。"医"的古文字是"醫"，本身就是一种酿造酒。古人酿酒目的之一是作药用的。远古的药酒大多是酿造成的，药物与酒醪混合发酵，在发酵过程中，药物成分不断溶出，才可充分利用。殷商时期的酒类，除了"酒""醴"之外，还有"鬯"。鬯是用黑黍为原料，加入郁金香草酿制而成，这是有文字记载的最早的药酒。

　　湖南长沙马王堆西汉汉墓出土的帛书《养生方》为一部医方专书，后来被称为《五十二病方》，被认为是公元前3世纪末的抄本，其

> Medicinal Wine

The "Food and Commodity Volume" of the book *History of the Han Dynasty* stated, "Wine is one of the best medicine inducers." In the remote ages, wine was regarded as a medicine, which was used to treat diseases. The ancient Chinese character for "medicine" is comprised of a part meaning wine. One of the purposes for people in ancient times to brew wine was to treat ailments. Ancient medicinal wine was always brewed by mixing and fermenting the medicine and the wine mash. During the fermentation process, the medicinal ingredients continuously dissolved into the liquid. Only then the medicinal ingredients became effective. There was a kind of wine called *Chang* in the Shang Dynasty, which was brewed from a mixture of black millet and tulip herbs. This was the earliest written record

中用到酒的药方不下于35个，至少有5方可认为是酒剂的配方。其中既有内服药酒，也有供外用的。

唐代著名医学家孙思邈所撰《千金要方》中列举了大量的药酒，他自己也是位养生专家。明代李时珍的《本草纲目》52卷中载有69种药酒，有的至今还在沿用。该书集明代及历代药物学、植物学之

of medicinal wine.

The silk book *On Health Care* excavated from tombs of the Western Han Dynasty in Mawangdui, Changsha City, Hunan Province was a book of prescriptions and also called *Prescriptions for Fifty-two Diseases*. The book was deemed as a manuscript written during the later years of the third Century B.C.. Among the thirty-five medical prescriptions involving medicinal wine, five were recipes of medicinal wine. Some of the medicinal wine were for oral administration and others for external application.

In his book *Comprehensive Clinical Medicine*, Sun Simiao, a famous medical scientist as well as a health specialist during the Tang Dynasty, listed many medicinal wine types. The fifty-two volumes on Chinese herbal medicine composed by Li Shizhen during the Ming Dynasty entitled *Compendium*

- **孙思邈像**

孙思邈，唐朝著名的医师，是中国古代著名的医学家和药物学家，被誉为"药王"。

A Portrait of Sun Simiao

Sun Simiao was a famous medical doctor. As a famous medical and farmaceutical scientist in ancient times, he was honored as the "King of Medicine".

• 李时珍像
A Portait of Li Shizhen

李时珍，中国古代伟大的医学家和药物学家，他历时27年编成的医学著作《本草纲目》是对中国古代药物学的总结。
Li Shizhen is a famous medical and pharamceutical scientist in ancient times. The book he compiled over twenty-seven years "*Compendium of Materia Medica*" was a summary of ancient Chinese pharmacology.

大成，广泛涉及食品学、营养学、化学等学科。该书的附方收集了大量的药酒配方，其卷二十五"酒"条下，设有"附诸药酒方"的专目，共计药酒方为200多种。

　　药酒是以白酒作为酒基，加入各种中药材，经过酿制或泡制而成的一种具有药用价值的酒。各种药酒因其用酒、用料的不同，酒精度也有不同；又因其加入的药料不

of Materia Medica recorded sixty-nine types of medicinal wine, some of which are still being used today. By integrating major achievements in the medicinal and the botanic sectors through many dynasties, the Compendium involved many sectors such as food, nutrition and chemistry. The attached prescriptions in the Compendium included recipes of medicinal wine prescribed. Under the article "wine" of chapter twenty-five of the book, there was a special provision of attachments with a total of more than two hundred medicinal wine prescriptions.

　　Using *Baijiu* as the base wine, Medicinal wine is made by mixing various traditional medicinal ingredients with wine. Then the mixture goes through further brewing or soaking processes and finally becomes wine of medical value. Due to differences in the wine type, raw materials, alcoholicity or medicines, the medicinal efficacy of these medicinal wines varies. High-quality medicinal wine uses high-quality *Baijiu* as the base wine, precious medicines and materials as the added ingredients for brewing and soaking. The juice is clear, transparent and pure, giving out a strong mixed aroma of the medicine and the wine, and has enduring aftertaste. Such wine is

同，其药用功效各有不同。优质药酒的特点是：以优质白酒为基酒，以名贵药材配制或浸泡，酒液清亮透澈，无杂质，药香、酒香和谐醇厚，回味无穷，对治疗疾病有辅助功效。

药酒品种繁多，功效各异。常见的药酒有五加皮酒、人参酒、灵芝酒、五味子酒、三蛇酒、虎骨酒、竹叶青酒等。中国配制的药酒在世界上享有盛誉。

conducive to treating diseases.

Medicinal wine has many varieties with different treatment efficacies. Popular Chinese medicinal wines include *Wujiapi*, Ginseng, *Lingzhi*, Wuweizi, *Sanshe*, *Hugu*, *Zhuyeqing*, etc. Chinese medicinal wine enjoys a high reputation in the world.

> 中国少数民族的特色酒

在中国少数民族中，有很多非常具有特色的酒。

蒙古族马奶酒：蒙古族传统的酿酒原料是马奶。马奶酒的酿制历史悠久，时至今日仍盛行于蒙古族牧区。马奶酒的酿制时间自夏伏骒马下驹时始，至秋草干枯马驹合群不再挤奶时止。这段时间被称为"马奶酒宴"期。

藏族青稞酒：青稞酒是藏族人民普遍喜爱的传统饮料，传说其酿制技术是7世纪唐代文成公主进藏传授的。在藏族聚居区，青稞酒是家家自酿，人人能饮的酒品。酒色黄绿清淡，酒味甘酸微甜，度数较低，10度左右；虽饮之难醉，但醉则难醒。藏族还另有青稞白酒，

> Special Types of Wine Made by Chinese Ethnic Groups

Many special types of wine are produced by Chinese ethnic groups.

Mongolian Koumiss Wine: Horse milk is the traditional raw material for making Koumiss by the Mongolian people in China. With a long brewing history, Koumiss still prevails in the Mongolian pastures. The brewing season for Koumiss commences in summer when mares deliver foals, and ends in late autumn when the pasture turns yellow and mares stop feeding, a season also known as the "Koumiss Festival".

Zang people's Highland-barley Wine: Highland barley wine is the traditional beverage among the Zang people. As legend has it, the technique for brewing this wine was brought by

酿制法较复杂，可称为"土法蒸馏"，酒精度可达60度以上，酒香四溢，略带青稞味。

柯尔克孜族孢糟酒："孢糟"是柯尔克孜语音译，可意译为"黄米酒"，因其原料是黄米。此酒酸甘相兼。

门巴族曼加酒："曼加"是藏语音译，意为"鸡爪谷酒"，因以当地特产的鸡爪谷为原料酿制而得名。饮用时，将发酵后的鸡爪谷（酒酿）装进底部有塞子的竹筒，加入凉水，稍候拔开塞子，以酒具接盛即可饮。曼加酒的度数仅有10度左右。

Princess Wencheng of the Tang Dynasty in the seventh Century when she married a King of Tubo. In the areas inhabited by Zang people, every household makes its own barley wine and everyone drinks it. The wine is light yellow or light green in color, clear and creatitine in texture, with a low alcohol content of ten percent. Although it is not so easy to get drunk on such low alcoholic beverage, it is equally difficult to sober up. Zang people also make Highland-barley *Baijiu* involving more complicated brewing processes, which could be called an "improvised

● 藏族青稞酒 (图片提供：全景正片)
Zang Barley Wine

● 铜青稞酒壶
A Bronze Barley Wine Lagon

水族九阡酒：九阡酒以糯米为主要原料，酿制过程中加入多种药材。酒色棕黄，状若稀释的蜂蜜，味微甘，酒香馥郁。九阡酒下窖的时间越长越醇。陈年九阡酒要在孩子出生时酿造下窖，直至其结婚时，甚至到寿终前才饮用。

土家族甜酒茶：土家族的甜酒茶实际上不是茶，而是酒。土家族以糯米或高粱煮甜酒，将甜酒和蜂蜜冲入盛了山泉水的碗中，甜酒茶

• 水族妇女用米酒迎接宾客 (图片提供：FOTOE)
Shui Women Offer Rice Wine to Welcome Guests

distilling technology". With a high alcoholicity of sixty percent, this wine gives out a strong aroma tinted with barley flavor.

Kirgiz people's *Baozao* Wine: The term "*Baozao*" is translated from the Kirgiz language, meaning "rice wine". Using rice as its raw material, *Baozao* wine has a sweet and sour texture.

Monba *Manjia* Wine: The term "*Manjia*" is translated from the Zang language, meaning "chicken feet millet wine". This wine is named after its raw material chicken feet millet, a local species. When drinking, people put the fermented chicken feet millet into a bamboo tube with a plug at the bottom, then infuse cold water. After a while, they just pull out the plug and pour wine into cups to drink. Manjia wine contains as little as ten percent alcohol only.

Shui people's *Jiuqian* Wine: This wine is made from sticky rice and some medicines. The color is brown as diluted honey with a sweet and aromatic texture. The longer it is stored underground, the stronger the aroma. Some *Jiuqian* wine is made and stored when a baby is born and is taken out for service only when he is getting married, or even when he is passing away.

即成。饮之清冽甜香，消暑提神。

普米族酥里玛酒：酥里玛酒主要以大麦和玉米为原料酿制。在密封的坛口插一支吸管，用酒时，以虹吸原理将酒引流出来。

普米族大麦黄酒：该酒的酿制是先将大麦煮熟，拌酒药发酵后，装入大土陶中，以灶灰泥封好坛口，21天后以管子引出酒液，装坛存放，随饮随取。该酒酒色橘黄，味道甘甜。

羌族蒸蒸酒：其酿制方法较简便，酒色淡黄，酒味甘甜，能去瘀血、生血下奶，是羌族产妇哺乳期间的常备饮品。

四川彝族苦荞酒：酿酒原料主要是苦荞（苦荞是凉山半高寒山区的特产，一种有清苦味的荞麦）、玉米或土豆；因主要以苦荞为原料酿制而得名。彝族人把用玉米、高粱和少量苦荞作原料酿制的酒叫作"泡水酒"。

云南彝族辣酒：辣酒的主要原料是玉米或高粱，待到发酵至白浆将要从竹箩孔中渗出时，装入坛中密封；当玉米或高粱变成细糊状时，装进甑子，上铁锅，兑水蒸馏出酒液。

Tujia people's Sweet-wine Tea: This sweet-wine tea is actually wine rather than tea. Tujia people brew sweet wine with sticky rice or sorghum. Then they infuse sweet wine and honey into the bowl with spring water. This is the sweet-wine tea. With a smooth, aromatic and sweet texture, the wine-tea is very refreshing.

Pumi people's *Sulima* Wine: This wine is made from brewed barley and corn. A straw is inserted into a sealed jar. When serving, the wine will flow out according to the siphon principle.

Pumi people's Barley Wine: This wine is made from barley which is first boiled until it's cooked. Then it is mixed with yeast to ferment and put into a big pottery jar and sealed with kitchen mortar. Three weeks later, people use a tube to extract the wine and store it in another jar for serving. This orange color wine has a sweet texture.

Qiang people's Steamed Wine: This light-yellow colored and sweet texture wine is very easy to make. As it could eliminate gore, generate more blood and breast milk, it is a very popular beverage for women during breast-feeding periods.

Sichuan Yi people's Buckwheat Wine: This wine is made of buckwheat in semi-alpine mountains of the Liangshan

怒族咕嘟酒：咕嘟酒的酿制法是将玉米粉制成酒。饮用时先将坛中的酒同酒糟盛一部分到盆中，加入适量开水，再拌入些蜂蜜或糖，滤去渣，饮其汁。

• 云南楚雄彝族烤酒
Yunnan Chuxiong Yi People's Baked Wine

Region, Sichuan Province, corn and potato. It is named after its major raw material buckwheat. The Yi people also use corn, sorghum and a small portion of buckwheat to make another type of wine called "Water-soaked Wine".

Yunnan Yi people's Chilly Wine: This is a wine brewed mainly from corn or sorghum. The fermented corn or sorghum is first put in a jar and sealed. When the corn or sorghum becomes a paste, it is put into a streamer on an iron boiler and filled with water. Then the wine juice will be distilled.

Nu people's *Gudu* Wine: This wine is made from corn flour. Prior to drinking, people put some wine and distiller's grain into a vessel, infuse some boiled water, honey or sugar, filter the lees and then drink the wine.

中国酒文化
Chinese Wine Culture

  酒文化是个大概念，与酒有关的思想道德和社会观念，可以说是与酒的酿造历史同期产生的。

The wine culture is a general concept. It maybe stated that the ideological, ethical and social concepts pertaining to wine emerged simultaneously with the history of wine-brewing.

> 古代的酒德和酒礼

历史上,儒家的学说被奉为治国安邦的正统观点,酒的习俗同样也受儒家酒文化观点的影响。儒家讲究"酒德"两字。"酒德"两字,最早见于《尚书》和《诗经》,其含义是说饮酒者要有德行,不能像商纣王那样,"颠覆厥德,荒湛于酒"(罔顾伦理道德,沉湎于酒色之中)。《尚书·酒诰》中集中体现了儒家的酒德,这

> Wine Ethics and Etiquettes in Ancient Times

In Chinese history, Confucian theory has been regarded as the orthodox theory governing the country and bringing peace and stability. Similarly, wine customs have also been influenced by Confucian views pertaining to the wine culture. Confucian was very meticulous about the "wine ethics" which was first seen in the book *Collection of Ancient Texts* and *The Book of Poetry*, meaning

- 战国铜器纹饰中的饮酒图
  The Drinking Scene as Decoration on Bronze Wares during the Warring States Period

就是："饮惟祀"（只有在祭祀时才能饮酒）；"无彝酒"（不要经常饮酒，平常少饮酒）；"执群饮"（禁止民众聚众饮酒）；"禁沉湎"（禁止饮酒过度）。儒家并不反对饮酒，认为用酒祭祀敬神，养老奉宾，都是德行。

饮酒作为一种食的文化，在远古时代就形成了大家必须遵守的、有时还非常繁琐的礼节，这就是酒礼。在古代祭祀时，酒是必不可少的祭品和礼节的组成部分。在重要的场合如果不遵守酒礼，就有犯上作乱的嫌疑；饮酒过量如不能自制，便容易生乱，故制定酒礼尤为重要。明代学者袁宏道看到酒徒不循酒礼，深感长辈有责任，便从古籍中采集了大量资料，专门写了一篇《觞政》，这虽然是为饮酒行令者写的，但对于一般的饮酒者也有意义。

具体说，古代饮酒的礼仪约有四步：拜、祭、啐、卒爵。就是先做出拜的动作，表示敬意，接着把酒倒出一点"酹（lèi）"在地上，以祭谢大地生养之德；然后尝尝酒味（啐）并加以赞扬令主人高兴；

that a drinker should have lofty moral standards and should not act as recklessly as King Zhou of the Shang Dynasty who indulged himself only in drinking and prurience. The chapter "Admonition on Wine" of the book *Collection of Ancient Texts* was a typical embodiment of Confucian wine ethnics which included texts such as: "People should not drink except when they are worshipping. They should not drink frequently. Group drinking is forbidden. Excessive drinking is forbidden". Confucian theory was not against drinking bluntly. Rather it believed in ethical ways to use the wine such as offering wine to gods and ancestors, nourishing the old and entertaining the guests.

As a component of the food culture, wine drinking has formed its own set of rules since ancient times. Drinkers must obey these rules which could be cumbersome at times. These were the wine etiquettes. In ancient China, wine was an indispensable item during worshipping rituals. On important occasions, if a person did not follow the drinking rules, he could be regarded as a suspect for riot. If a person overdrank, it was easy for him to lose self-control and this could invite trouble. Here wine

最后卒爵仰杯而尽。

　　古代酒礼内容丰富，许多饮酒礼制甚至影响数千年，并成为东方餐饮文化的重要组成部分。比如：同座会饮叫"宴"；能够饮酒的人饮酒，不会饮酒的陪坐叫"醧（yù）"。主客相互敬酒叫"酬酢"。闭门饮酒，并久而不止叫"湎"。饮酒成瘾，不能自拔，叫"沉"。

　　礼制上规定：君子可以宴，可以醧，不可沉，不可湎。

　　酒礼还包括斟酒、敬酒、祝酒、饮酒方式等。在酒宴上，主人向客人敬酒，客人要回敬主人，敬酒时还要说上几句敬酒辞。客人之间相互也可敬酒（称"旅酬"）。有时还要依次敬酒（称"行酒"）。敬酒时，敬酒的人和被敬酒的人都要"避席"，即起立。普通敬酒以三杯为度。

　　中国古代饮酒还有一些礼节。如主人和宾客一起饮酒时，要相互跪拜。晚辈在长辈面前饮酒，叫"侍饮"，通常要先行跪拜礼，然后坐入次席。长辈命晚辈饮酒，晚辈才可举杯；长辈酒杯中的酒尚未饮完，晚辈也不能先饮尽。

etiquette played an important role. When Yuan Hongdao, a famous Ming-dynasty scholar saw drinkers were violating the wine etiquettes, he felt it his obligation to guide these young people. So, he wrote an article entitled *The Drinking Policy* (*Shang Zheng*) after consulting many classic books and documentation. Although the article was catered mainly towards those wine drinkers who played line games while drinking, it also bore reference value for ordinary wine drinkers.

　　Specifically, ancient drinking etiquettes entailed four steps. First, the drinker bowed to show respect, then sprinkled a drop on the ground as an offering to Mother Earth to express his gratitude. Then he took a small sip and praised the wine to the host. Finally, he drank the whole cup in one go.

　　Wine etiquettes had rich connotations in ancient times; many of them had far-reaching influence and had become an important component of the oriental catering culture. For example, special terminologies have been developed for drinking together around a table, for those who are not drinking, for toasting between the hosts and the guests, for drinking alone, and for alcohol addicts.

　　The wine etiquettes stipulated that

a gentleman is allowed to drink with others, or accompany others to drink, but he should not indulge in over-drinking or become an alcoholist.

The wine etiquettes also include proper manners during the whole course of the banquet, for instance, how to fill the cups, how to propose a toast and how to drink a toast. In a banquet, the host should toast the guests, and the guests should also return the host. The toasts should be accompanied by some simple message of appreciation or gratitude. The guests may also propose toasts among themselves. Sometimes, people need to toast around the table. When toasting, both the toaster and the toasted should stand up. Generally speaking, one round of toast should not be more than three cups of wine.

There were also some other wine etiquettes in ancient China. For instance, when the hosts and the guests drink together, they should bow on their knees to each other. When drinking together with seniors, the juniors should first bow on their knees and then sit in secondary seats. The juniors should not start drinking until told by the seniors to do so and they should not finish the wine in their cups before the seniors did.

饮酒讲究礼节
Wine Etiquette

> 酒令

　　酒令、酒筹是筵宴上助兴取乐的饮酒游戏，最早诞生于西周，完备于隋唐。一般是席间推举一人为令官，余者听令或赋诗填词、对联语，或猜谜行拳做其他游戏，输者或违反规则者要被罚饮酒，故又称"行令饮酒"。这需要行酒令者敏捷机智，有文采和才华。因此，饮酒行令既是古人好客传统的表现，又是他们饮酒艺术与聪明才智的结晶。

　　行令饮酒在古代的士大夫中特别风行，他们还常常赋诗撰文予以赞颂。行酒令的方式可谓是五花八门。酒令包括雅令、通令等方式。雅令的行令方法是：令官或出诗句，或出对子，其他人按首令之意续令，所续必在内容与形式上与首令相符，不然则被罚饮酒。通令

> **Drinking Games**

Drinking games first appeared during the Western Zhou Dynasty, and matured during the Sui and Tang dynasties. The common practice was that one person would be nominated as the head of the table, while the rest followed his instructions to compose a poem, a verse, a couplet, or guess riddles or play finger-guessing games. If a person failed the game or violated the rules, he would be punished for drinking wine. These games required the participants to be wise, clever, quick and knowledgeable. Therefore, playing games while drinking wine was not only an expression of the hospitality of the host, but also reflected their wisdom and drinking style.

　　Playing games while drinking was very popular among ancient literati, who often composed poems and articles to

的行令方法主要是掷骰、抽签、划拳、猜数等。

不过，文人雅士与平民百姓行酒令的方式大相径庭。文人雅士常用对诗或对对联、猜字或猜谜等方式行令，而一般百姓则用些简单通俗又不需作任何准备的行令方式，如二人同时喊出"老虎、杠子、鸡"其中之一，则"杠胜虎、虎胜鸡、鸡胜杠"。总的说来，酒令是用来罚酒，但其最主要的目的是活跃饮酒时的气氛。另外，酒令还有一个功能，就是古代礼仪教化的方式之一，故能盛行于各朝各代。

古人还有一种"投壶"的饮酒习俗，源于西周时期的射礼。射礼为宴饮而设，称为"燕射"。酒宴上设一壶，宾客依次将箭向壶内投去，以投入壶内多者为胜，负者受罚饮酒。

中国人饮酒最讲究的是意境。最佳的状态是似醉非醉，所谓"花看半开，酒饮微醺"，使人处于最快乐、最迷蒙的境界中。

record and praise the occasions. There were various forms of drinking games; some were elegant while others were popular. In the case of an elegant game, the head of the table would take the lead in composing a poem; other participants would follow the theme and the style, and complete the poem. Those who failed to come up with appropriate lines would be punished by drinking more wine. Popular games included casting dice, drawing straws, guessing fingers, guessing numbers, etc.

Literati's drinking games were obviously different from that of the rank-and-file. The games usually involved poem composition, character and riddle guessing. The rank-and-file's games tended to be simpler and easy to play. For example, when two people cried out any of the "tiger, bar and rooster" simultaneously, the bar would defeat the tiger, the tiger defeat the rooster and the rooster defeat the bar. Generally speaking, drinking games were aimed at punishing the loser to drink more wine. However, the most important purpose of the game was to cheer up the atmosphere. Another function of the games was to disseminate the ancient wine ethics and etiquettes. That's why drinking games

• 陶投壶（汉）
A Pottery Cast Pot（Han Dynasty, 206 B.C.-220 A.D.）

have been very popular through many dynasties.

Ancient Chinese people also played a casting game, which had its origin in the shooting ceremony during the Western Zhou Dynasty. The game was specifically designed for wine feasts. During the banquet, a wine pot would be placed in the middle and all the guests were invited to cast arrows into the pot. The one who got the most arrows in would be the winner. The loser would be punished by drinking wine.

Ancient Chinese people cared more about the ambience and the mood accompanying the drinking. The best state for a drinker was when "he is half drunk and half sober, like a half-blooming flower". This was believed to be the happiest and most exciting moment for the drinker.

> 酒与民俗

在中国古代，酒的使用是庄严之事，非祀天地、祭宗庙、奉佳宾而不用。随酿酒业的普遍兴起，远古酒事活动的俗尚和风格形成后，酒逐渐成为人们日常生活的用物，酒事活动也随之广泛，逐渐演变为系统的酒风俗习惯。这些酒俗涉及人们生产、生活的许多方面。

酒与节日

中国是世界上节日最多的国家之一。从古到今，中国人在一年中的几个重大节日中，都有相应的饮酒活动。

> Wine and Folk Customs

In ancient times, wine was only used for important occasions such as worshipping the heaven and the earth, commemorating ancestors, or entertaining distinguished guests. With the general rise of the wine industry, wine gradually became people's daily beverage. Activities involving wine became more widespread and wine customs more systematic, involving many aspects of people's production activities and daily life.

Wine and Festivals

China is one of the countries celebrating the most festivals in the world. Since ancient times, Chinese people cultivated the tradition of holding drinking parties to celebrate several major festivals throughout the year.

春节

春节俗称"过年"。春节期间要饮用以屠苏泡的屠苏酒、以椒柏泡的椒花酒，辟疫疠及一切不正之气，寓意吉祥、康宁、长寿。宋代王安石在《元日》一诗中写道："爆竹声中一岁除，春风送暖入屠苏。千门万户瞳瞳日，总把新桃换旧符。"

元宵节

元宵节又称"灯节""上元节"。唐代起，元宵张灯成为法定之事，过去人们向天官祈福，必用五牲、果品、酒供祭。祭礼后，家人团聚畅饮一番，以祝贺新春佳节结束。

The Spring Festival

The Spring Festival is the New Year for the Chinese people. During the Spring Festival, people should drink the *Tusu* herbal wine and the Pepper wine, so as to avoid plagues and evils. It also symbolized good luck, good health and longevity. In his poem *The New Year's Day*, Song-dynasty poet Wang Anshi wrote: "The old year is gone in resounding firecrackers. People drink *Tusu* wine bathed in warm spring breeze. The morning sun shines over all the households and people are busy replacing the old door god with a new Divinity."

The *Yuanxiao* Festival

The *Yuanxiao* Festival is also known as the "Lantern Festival" or the "*Shangyuan* Festival". Since the Tang Dynasty, the custom of decorating with lanterns during the *Yuanxiao* Festival has become a legal requirement, and people usually prayed to the Heaven for happiness. They usually used five animals (buffalo, sheep, swine, dog and chicken), fruits and wine as offerings. After the ceremony, the whole

- 元宵节，赏花灯 (图片提供：全景正片)
Enjoying the Beautiful Lanterns during the *Yuanxiao* Festival

中和节

又称"社日节",时在农历二月初一,随着历史的变化,改为二月初二。据文献记载:"村舍作中和酒,祭勾芒种,以祈年谷。"这天有饮中和酒、宜春酒的习俗,以祭祀土神,祈求丰收。据说这天饮酒还可以医治耳疾,因而人们又称之为"治聋酒"。

清明节

寒食节与清明节合为一个节日,时间在每年阳历4月4日至6日之间变动。清明节有扫墓、踏青的习

- 牧童遥指杏花村
  The Young Cowboy Pointing to the Distant Apricot Blossom Village

family would sit together to enjoy a feast of wine, marking the completion of the New Year season.

The *Zhonghe* Festival

The *Zhonghe* Festival, also known as the "Earth Day", falls on the first day of the second lunar month, and has changed to the second day of the second lunar month over the course of history. According to ancient documentation, people would make *Zhonghe* wine or *Yichun* wine on that day, and offer to the Earth God, to pray for bumper harvests. Because it was said that the wine could cure ear disorders, the wine was also called the "Deaf-curing Wine".

The Tomb-sweeping Day

The Tomb-sweeping Day (merged with the Cold Food Day) usually falls on April 4th, 5th or 6th according to the Gregorian Calendar, on the day of Pure Brightness in solar terms. Because people would usually go to the graveyard to commemorate the deceased or go on an excursion in the countryside, drinking was not prohibited on this day. Reasons for allowing people to drink wine on this occasion were two folds: a) since only cold food would be served on that day, and wine could warm people up

俗，此时饮酒可不受限制。清明节饮酒，一是寒食节期间只能吃凉食，饮酒可以增加热量；二是借酒来平缓人们哀悼亲人的心情。古人为清明饮酒赋诗较多，唐代诗人杜牧在《清明》一诗中写道："清明时节雨纷纷，路上行人欲断魂。借问酒家何处有？牧童遥指杏花村。"

端午节

农历五月五日又称"端阳节""重午节""重五节"等。唐代光启年间即有饮"菖蒲酒"的实例，人们以此辟邪、除恶、解毒。菖蒲酒是以菖蒲泡成，是传统的时令饮料，历代帝王也将它列为御膳时令香醪。明代刘若愚在《明宫史》中记载："初五日午时，饮朱砂、雄黄、菖蒲酒，吃粽子。"

中秋节

中秋节又称"仲秋节""团圆节"，时在农历八月十五。在这个节日里，无论家人团聚，还是挚友相会，都离不开赏月饮酒。清代以后，中秋节以饮桂花泡制的桂花酒为习俗。

and increase the calorie; b) drinking could help people feel relieved of their deep sorrow for the dead. There were many poems composed by ancient Chinese poets for this day. In his poem *Qingming*, Du Mu, a famous poet of the Tang Dynasty wrote the following well-known verses: "It's raining on Pure Brightness Day, making people going to the graveyard to commemorate the dead all the more sorrowful. When a passer-by asked a young cowboy where he could find a wine restaurant? The boy points to a distant village beyond the apricot blossom village."

The Loong-boat Festival

The Loong-boat Festival, also called the "*Duanyang* Festival" or the "*Chongwu* Festival", falls on the fifth day of the fifth lunar month. During the Guangqi Period (885-888) of the Tang Dynasty, people developed the tradition of drinking "Calamus wine", in order to expel evils, ghosts or toxic matters. Calamus wine was a traditional beverage designated by past Chinese emperors as a seasonal imperial drink. In his book *History of the Ming Palaces*, Liu Ruoyu wrote: "At noontime on the fifth day of the fifth lunar month, it is time to take cinnabar

重阳节

重阳节又称"重九节""登高节",时在农历九月初九,登高饮酒的习俗普及于西汉。历代人们逢重九就要登高、赏菊、饮酒,延续至今不衰。除饮菊花酒(以菊花泡制)外,有的还饮用以茱萸泡制的茱萸酒和以茱萸、菊花泡制的茱菊酒,以及黄花酒、薏苡酒、桑落酒、桂酒等。

● 菊花
Chrisanthamums

and realgar, drink calamus wine and eat *Zongzi*, which is a pastry made of glutinous rice wrapped in weed leaves."

The Mid-Autumn Festival

The Mid-Autumn Festival, also known as the "*Zhong-Qiu* Festival" or the "Reunion Festival", falls on the fifteenth day of the eighth lunar month. This is the time when wine would be served at family reunions or gatherings of friends. People would enjoy the full moon while sipping delicious wine. Since the Qing Dynasty, people developed the tradition of drinking osmanthus wine during this festival.

The Double Ninth Festival

The Double Ninth Festival also known as the "Mountain-climbing Festival", falls on the ninth day of the ninth lunar month. The festival tradition became popular since the Western Han Dynasty. People would climb to the top of the mountain and drink wine while enjoying chrysanthemums and the beautiful landscape. In addition to chrysanthemum wine, comel wine, chrysanthemum and comel mixed wine as well as lily wine, coix-seed wine, mulberry wine and osmanthus wine were also popular drinks on this occasion.

除夕

除夕即阴历大年三十之夜，有别岁、守岁的习俗，即除夕夜通宵不寐，回顾过去，展望未

New Year's Eve

New Year's Eve is the last night of the year when people would often stay up late or even the whole night to re ect on the past year and to welcome the New Year. On this occasion, people used to drink wine. In his poem *Stay up all night on New Year's Eve*, Bai Juyi, a famous poet in the Tang Dynasty, wrote: "At that night, we drink much wine until the jars are empty. We are all homesick that tears wet the handkerchief." The dinner of that evening is usually a family feast when the young generation of the family toasts the old. The custom has been extended until modern times.

Ethnic Groups Festivals

In addition to the above festivals, there are also many festivals observed by ethnic groups that take drinking and entertainment as one of the major activities. For instance, the Bathing Festival and the Fruits Festival observed by Zang people; the Sisters Festival and the Begging for Sapling Festival of Miao

- 云南民族村敬酒台（图片提供：全景正片）
The Toasting Balcony in Yunnan Ethnic Group Village

来。除夕守岁都是要饮酒的，唐代诗人白居易在《客中守岁》一诗中写道："守岁樽无酒，思乡泪满巾。"除夕夜一家人的聚餐又名为"团圆酒"，要向长辈敬辞岁酒，这一习俗延续到今。

民族节日

以饮酒游乐为主要活动的民族节日亦不少，如藏族的沐浴节、望果节，苗族的姊妹节、讨树苗节，锡伯族的西迁节，朝鲜族的流头节，蒙古族的马奶节，等等。另外，在民族节日中，人们往往把酒当作礼品或奖品。姑娘们酿的酒或送的酒也叫"姑娘酒"。若是在摔跤、赛马、斗牛等各项比赛之后，姑娘们喜欢得胜者，也会拥上前去给他们敬酒，表示赞赏和敬慕。在这种交往中结为良缘的亦有不少。

酒与习俗

中国有众多民俗活动，酒都在其中扮演着重要角色。农事节庆时的祭拜庆典需有酒，缅怀先祖、追求丰收富裕的情感就得以寄托；婚嫁需有酒，白头偕老、忠贞不渝的爱情得以明誓；丧葬需有酒，后人忠

people; the Westward Migration Festival of Xibo people; the Bathing Festival of the China's Korean ethnic group; the Koumiss Festival of the Mongolians; etc. During these festivals, local people usually use wine as gifts or prizes. Wine made by girls or presented by girls is called "Girls' Wine". In events such as a wrestling competition, a horse race, a bullfight or other contests, girls usually admire the winner. They will rush to the hero to toast him. It is quite common for young people to fall in love during such events.

Wine and Customs

Wine plays an important role in many folk activities. It seems that a wine-free worshipping ritual, a commemoration ceremony or a praying session for good harvests would have no carrier for people's sentiments; a wine-free wedding ceremony would have no vehicle for the bride and the groom to vow for their mutual love; similarly, a wine-free funeral would make the descendants feel nowhere to demonstrate their grievance for their beloved ancestors; a wine-free birthday party would be difficult to show people's etiquettes and interests; a wine-free farewell party for soldiers

孝之心得以表述；生宴需有酒，人生礼趣得以显示；饯行洗尘需有酒，壮士一去不复返的悲壮情怀得以倾诉。总之，无酒不成俗。

与习俗相关的酒主要有以下几类：

绍兴花雕酒
Shaoxing Carved Wine

going off to the battlefield would be difficult to express the heroism and the tragic feelings for those leaving. In short, customs would not establish themselves without wine.

The main types of wine related to folk customs are the following.

Daughter's Wine: According to the book *Grass and Trees in Southern China — A Monograph on Botany* written by Ji Han of the Jin Dynasty (304), if a family in southern China gave birth to a baby girl, the parents would brew wine and bury it at the bottom of a lake when the daughter was still very young. The wine would be taken out to entertain the guests when the daughter got married. The reason for adopting such a brand name is because, on the pottery jars containing this wine, there are usually carved auspicious patterns such as flowers, birds, animals, figures, mountains and pavilions. When the daughter gets married, the jars are taken out and color patterns would be added to the carvings, just as a blessing for good luck.

Happy-union Wine: In ancient times, the bride and the groom would exchange a joint drink of wine from two scoops that were made from the same

女儿酒：晋代的嵇含所著《南方草木状》说，南方人生女仅数岁，便开始酿酒，酿成酒后，埋藏于池塘底部，待女儿出嫁之时才取出供宾客饮用。这种酒在绍兴得到继承，发展成为著名的"花雕酒"。盛放女儿酒的酒坛在制坯时，就雕上各种花卉、人物鸟兽、山水亭榭等吉祥图案；等到女儿出嫁时，取出酒坛，请画匠再用油彩添上色，以讨好彩头。

合欢交杯酒：在古代又称为gourd, metaphorically implying that the new couple belongs to each other. This practice was further evolved during the Song Dynasty when two cups would be linked with a piece of colorful silk thread and tied into a heart-shaped knot. The new couple would drink the connected cups or from the same cup in turns, symbolizing that they would be as one.

Cross-arm Wine: In order to show their mutual love, the new couple is requested to cross their arms while holding a wine cup. They first stare into

• 合卺杯（明）
Happy-union Wine Cup (Ming Dynasty, 1368-1644)

"合卺"（jǐn，卺即瓠分成的两个瓢），古语有"合卺而酳"，即新娘新郎各执一片卺以酳(xǔ，即以酒漱口)，"合卺"取"你中有我，我中有你"之意，引申为结婚的意思。在唐代即有"交杯酒"这一名称，到了宋代，在礼仪上盛行用彩丝将两只酒杯相连，并绾成同心结之类的彩结，夫妻互饮一盏，或夫妻传饮。

交臂酒：为表示夫妻相爱，在婚礼上夫妻各执一杯酒，手臂相交，双目对视，各饮一口，实际上成了"交臂酒"。

月米酒：妇女分娩前几天，要煮米酒一坛，一是为分娩女子坐月子催奶，二是准备款待前来贺喜的近亲。

满月酒、百日酒：孩子在满月或百日时，父母要摆上几桌酒席，邀请亲朋好友共贺。亲朋好友一般都要送上礼物或红包。

寿酒：中国人有给老人祝寿的习俗，五十岁以上的老人过整岁生日，被称为"大寿"，一般由儿女或者孙辈出面举办，邀请亲朋好友参加酒宴。

祭拜酒：逢年过节或遇灾有难

each other's eyes and then drink the wine.

Maternity Rice Wine: A few days before a woman delivers a baby, her family usually prepares a jar of Rice wine which will be used to help the young mother produce more milk, and to entertain guests who might come to congratulate the birth of the new baby.

One-month-birthday or One-hundred-day-birthday Wine: When a baby is one month old or one hundred days old, his parents will throw a party to celebrate, inviting relatives and friends. The relatives and friends always bring gifts or red packets to the family.

Birthday Wine: It is a common practice to celebrate the major birthdays of old people when they reach the age of fifty, sixty, seventy, etc. Such parties are usually organized by the senior person's children or grandchildren. Relatives and friends are invited to participate in such parties where delicious food and wine are served.

Worshipping Wine: This is the wine for all kinds of rituals during festivals, or in cases of difficulties or disasters. On New Year's eve, all families would prepare sumptuous food and wine, burn joss sticks, candles and paper money. During the ritual, all family members

• 坛装米酒
Rice Wine in a Jar

would kowtow in front of the image of ancestors according to their seniority and then stand silently beside the offering table. After that, the most senior member of the family would sprinkle some wine on all sides of the table to complete the worshipping ceremony.

Foundation-laying, Beam-installing and House Completion Wine: Home building is an important activity for rural families. As foundation laying and beam installing are the two most important steps in building a house, the owner usually holds a wine-drinking ceremony to mark the occasion. They sprinkle wine on the ground as offerings to the God of Earth, and wine on the beam. When the new house is completed, another wine party is given to celebrate the move-in and to pray to ancestors for their blessings.

Business Launching and Dividend Distribution Wine: These are celebration wine parties when a new shop or a mill

时，要设祭拜酒。除夕夜，各家各户要准备丰盛酒菜，燃香点烛化纸钱，此间，以长幼次序磕头，随即肃穆立候于桌边，家长将所敬之酒洒于餐桌四周，祭拜才算结束。

奠基酒、上梁酒和进屋酒：在农村盖房是件大事，而奠基与上梁是盖房最重要的工序，故到了这时要办奠基酒或上梁酒，还要用酒祭地浇梁。等房子造好，举家迁入新居时，还要办进屋酒，一是庆贺新屋落成，并志乔迁之喜；二是祭祀神仙祖宗，以求保佑。

开业酒和分红酒：这是店铺作坊置办的喜庆酒。店铺作坊开张、开工之时，老板要置办酒席，以志喜庆贺。店铺作坊年终按股份分配红利时，要办"分红酒"。

壮行酒（送行酒）：有朋友远行，为其举办酒宴，表达惜别之情。勇士上战场执行重大且有生命危险的任务时，指挥官都会为他们斟上一杯酒，用酒为勇士们壮胆送行。

starts the business operation. At the end of the year when the shop or the mill distributes dividends to the shareholders, the owner usually holds a drinking party to celebrate the success of the business.

Farewell Wine or Seeing-off Drink: When someone is going on a long journey, his friends usually hold a wine party to say goodbye to him. Before warriors go to the battlefield, or set off on a dangerous mission, the commander usually offers them a drink of wine as a sign of encouragement.

> 酒与唐诗宋词

　　诗言志，酒载情。在历代文人骚客的诗词中，时时可见借酒言志抒怀的词句。

　　唐宋诗词是中国古代文学的一个发展高峰，唐朝人把饮酒赋诗当作最好的消遣娱乐方式之一，用"诗酒人生"来概括唐代文人的生活毫不夸张。其中杰出的代表人物就是李白与杜甫。

　　李白，字太白，是唐代最伟大的诗人之一。李白善饮，与他同时代的杜甫在《饮中八仙歌》写道："李白斗酒诗百篇，长安市上酒家眠。天子呼来不上船，自称臣是酒中仙。" 这四句诗，一是写出了诗人与酒的密切关系，二是写出了诗人同市井平民的亲近，三是写

> Wine and the Tang Poems and Song *Ci*-poems

Poetry expresses one's ambition while wine carries one's passion. In the poems and verses of many Chinese poets and literati, there were many wine-related verses that express the poet's ambition and aspiration.

　　Tang poems and Song *Ci*-poems were at development peak of classical Chinese literature. Tang people regarded composing poems while drinking wine as one of the best forms of entertainment. It cannot be more appropriate to describe the life of Tang literati as "life of poems and wine". Among them, Li Bai and Du Fu were the two most typical representatives.

　　Li Bai, courtesy name Li Taibai, was one of the greatest poets of the Tang Dynasty. Li Bai loved drinking. In

• 李白像

李白，唐朝伟大的浪漫主义诗人。他的存世诗文多达千余篇，其中与酒有关的代表作是《月下独酌》和《将进酒》。

A Portrait of Li Bai

Li Bai, was the great romantic poet of the Tang Dynasty. He wrote more than one thousand poems during his lifetime. Among them *Lonely Drinking under the Moon Light* and *The Toasting Song* were the two typical poems related to wine.

出了诗人孤高自傲的性情。故此，百姓都很喜欢李白，称他为"诗仙""酒仙"。为了称颂和怀念这位伟大的诗人，在古时的酒店里，都挂着"太白遗风""太白世家"的招牌，此风流传到近代。

与李白齐名的杜甫，"性豪业嗜酒，嫉恶怀刚肠"，多有"沉饮聊自遣，放歌破愁绝"的诗作。他

his poem *Songs for the Eight Drinking Immortals*, Du Fu wrote: "Li Bai could compose many poems after drinking wine. He always slept in wine restaurants in Chang'an City. When the emperor called him to board a boat, he responded slowly and said he was a wine immortal." This four-line poem has three meanings: a) Li Bai was closely related to wine; b) he was very intimate with the rank and file; c) he was independent and self-confident. That's why the common people loved him. They cordially called him the "immortal poet" or "the immortal wine drinker". In order to praise and commemorate this great poet, many Chinese restaurants and hotels hung a signboard "Taibai Spirit" or "Taibai Style". This tradition has been extended until modern times.

An equally famous poet Du Fu who was described as someone who "loved drinking and hated injustice" also wrote many unique poems expressing his "indulgence in drinking to expel his frustration, and his poems to voice his indignation". Many of his poems were about his concerns over the people and the state. One of the famous "wine poems" reads: "Aroma of meat and wine emanates from rich houses while the poor die in the street of starvation and cold."

的酒诗中虽然不乏娱情放纵、借酒浇愁的内容，但更多的是关注苍生社稷。杜甫有名的"酒诗"是"朱门酒肉臭，路有冻死骨。"（富贵人家门前飘出酒肉的味道，穷人们却在街头因冻饿而死。形容贫富悬殊的社会现象）。诗句散发着忧国忧民情怀。

This poem vividly depicted the wealth disparity between the rich and the poor, and fully expressed the poet's concerns over the people and the state.

- 杜甫像

杜甫，字子美，唐朝伟大的现实主义诗人，被后人尊称为"诗圣"，与李白合称"李杜"。

A Portrait of Du Fu

Du Fu, courtesy name Zimei, was a great realistic poet of the Tang Dynasty. Du Fu was respectifully known as the "Poet Saint" by later generations. His name was very often juxtaposed with another famous poet Li Bai.

- 《太白醉酒图》苏六朋（清）

此图描绘的是李白醉酒于唐玄宗宫殿之内，由内侍二人搀扶侍候的情景。

Inebriated Li Taibai, by Su Liupeng (Qing Dynasty, 1616-1911)

The painting depicts the inebriated Li Bai being attended by two servants in the Palace of Emperor Xuanzong of the Tang Dynasty.

盛唐饮中八仙长安酒会

饮中八仙是指唐朝嗜酒的八位学者名人,也称"酒中八仙"或"醉八仙"。他们分别是"诗仙"李白、诗人贺知章、汝阳王李琎、左相李适之、书法家张旭、辩论高手焦遂、美少年崔宗之和素食主义者苏晋。

盛唐时期盛行各种酒会,参与者甚众。历史上并没有"饮中八仙"齐聚一堂的明确记载,但在杜甫的诗《饮中八仙歌》描写了八仙聚饮的情景。诗人用精练的语言"勾画"了一幅栩栩如生的"群像画",展现了各具特色的八个人物。

饮中八仙歌

杜甫(唐)

知章骑马似乘船,眼花落井水底眠。
汝阳三斗始朝天,道逢麹车口流涎,恨不移封向酒泉。
左相日兴费万钱,饮如长鲸吸百川,衔杯乐圣称避贤。
宗之潇洒美少年,举觞白眼望青天,皎如玉树临风前。
苏晋长斋绣佛前,醉中往往爱逃禅。
李白一斗诗百篇,长安市上酒家眠。
天子呼来不上船,自称臣是酒中仙。
张旭三杯草圣传,脱帽露顶王公前,挥毫落纸如云烟。
焦遂五斗方卓然,高谈雄辩惊四筵。

译文:

贺知章酒后骑马,晃晃悠悠如在乘船。他眼睛昏花坠入井中,竟在井底睡着了。汝阳王李琎饮酒三斗以后才去觐见天子,路上碰到装载酒曲的车,酒味引得他口水直流,为自己没能被封在水味如酒的酒泉郡而遗憾。左相李适之为每日饮酒不惜花费万钱,饮酒如长鲸吞吸百川之水,自称举杯豪饮是为了脱略政事,以便让贤。崔宗之是一个潇洒的美少年,举杯饮酒时,常常傲视青天,俊美之姿有如玉树临风。苏晋虽在佛前斋戒吃素,饮起酒来常把佛门戒律忘得干干净净。李白饮酒十斗,立可赋诗百篇。他去长安街酒肆饮酒,常常醉眠于酒家。天子在湖池游宴,召他为诗作序,他因酒醉不肯上船,自称是酒中之仙。张旭饮酒三杯,即挥毫作书,时人称为"草圣"。他常不拘小节,在王公贵戚面前脱帽露顶,挥笔疾书,若得神助，其书如云烟之泻于纸张。焦遂五杯酒下肚，才得精神振奋，在酒席上高谈阔论，常常语惊四座。

Eight Drinking Immortals Meeting in Chang'an City during Flourishing Tang Period

The eight drinking immortals refer to eight literati celebrities during the Tang Dynasty. They were also known as the "Eight Wine Immortals" or the "Eight Drunken Immortals". They were poet Li Bai, poet He Zhizhang, King of Ruyang Li Jin, prime minister Li Shizhi, calligrapher Zhang Xu, articulate debater Jiao Sui, smart youngster Cui Zongzhi and vegetarian Su Jin.

The Flourishing Tang Period witnessed various wine parties with many attendees. However, there was no specific record of a party where all the eight immortals were present. Nevertheless, in his poem *Songs for the Eight Drinking Immortals*, Du Fu depicted the scene of the eight immortals' gathering. Using simple and precise language, Du Fu portrayed a vivid "group picture" of the eight drinking immortals, with each showing his unique characters.

- 《古贤诗意图》之《饮中八仙》杜堇（明）

  《饮中八仙》描绘的也是李白、贺知章等八人醉饮的场景，人物神态各异，场面热闹。

  Eight Drinking Immortals of the *Collection of Paintings and Poetry of Ancient Talents* by Du Jin (Ming Dynasty, 1368-1644)

  The *Eight Drinking Immortals* of the Collection of Paintings and Poetry of Ancient Talents also depicts the eight drinking literati drinking to their heart's content.

Songs for the Eight Drinking Immortals
by Du Fu (Tang Dynasty)

After drinking, He Zhizhang rode on horseback which was as shaky as in a boat. He became dim-sighted and fell into a well where he fell asleep.

Li Jin (King of Ruyang) went to see the emperor after he drank a lot of wine. On the way to the Imperial Palace, he saw a cart carrying wine yeasts. The strong aroma made his mouth watered. He regretted so much that he was not appointed as the governor of Jiuquan City where many good wines were produced.

Li Shizhi (a prime minister) spent much money on buying good wine, and he drank wine like a wale drinking water every day. He said the reason for his drinking was because he wanted to get rid of the heavy workload on him, so that new talents could come to replace him.

Cui Zongzhi was a handsome young man. When he drank wine, he often stared into the sky. At that moment he looked like a graceful jade tree facing the wind.

Although Su Jin was a vegetarian in front of Buddha, he often forgot Buddhist rules and taboos when he drank wine.

Li Bai could compose one hundred poems after he drank ten *Dou* (a measurement unit) of wine. He always slept in wine restaurants in Chang'an City. When the emperor called him to join him on a boat trip and to write a preface for his poems, Li Bai was so drunk that he refused to get on board mumbling that he was a wine immortal.

After drinking three cups of wine, Zhang Xu could immediately write beautiful calligraphy works. So he was nick-named "A Calligraphy Master". Being a sloppy person, Zhang Xu paid little attention to his attire and manner. He would take off his hat, reveal his head, and start writing feverishly in front of senior officials or members of the imperial family. His calligraphy was so forceful and fluent like clouds pouring on paper, as if he were assisted by God.

Jiao Sui was not active and talkative until he drank five cups of wine. Then, he became so excited that he would come up with many thought-provoking comments that stunned others around the table.

- 《醉饮图》【局部】万邦治（明）

The Drinking Scene [Part] by Wan Bangzhi (Ming Dynasty, 1368-1644)

《醉饮图》描绘的是当时的八位豪饮雅士，即李白、贺知章、李琎、李适之、张旭、焦遂、崔宗之和苏晋，他们开怀痛饮，各具醉姿，生动传神。

The painting depicts the drunken gesture of the eight famous drinker literati, i.e. Li Bai, He Zhizhang, Li Jin, Li Shizhi, Zhang Xu, Jiao Sui, Cui Zongzhi and Su Jin.

唐代同样著名的诗人白居易一生作诗三千多首，其中写到酒的就有好几百首。他有许多诗的题目就和酒有关，或者全诗都是写酒的，如《南亭对酒送春》《劝酒寄元九》《对酒》《花下对酒二首》《醉歌》《同韩侍郎游郑家池吟诗小饮》《醉后走笔》《醉后狂言》《同崔存度醉后作》《同李十一醉忆元九》《花下自劝酒》《答劝酒》《强酒》《春酒初熟》《醉吟二首》等，不一一列举。他最著名的长诗之一《琵琶行》就是他边喝酒边在琵琶声中酝酿创作而成的。白居易一生不仅以狂饮著称，而且以善酿出名。他为官时，分出相当一部分精力去研究酒的酿造。在酿酒的过程中，他不是发号施令，而是亲自参加实践。

灯红酒绿，处处笙歌，宋代人市井生活丰富。文人们在推杯换盏之际，酒酣兴浓之时，创作出了大量的诗词，宋人的好酒在宋词里被表现得淋漓尽致。北宋文学家苏轼的一首词就是《水调歌头》，给人一种非常浪漫、十分潇洒的感觉。

大文学家欧阳修是妇孺皆知的

Another equally famous poet Bai Juyi of the Tang Dynasty composed more than 3000 poems during his lifetime. Among them several hundred were related to wine. Some had wine in the title. Others were about wine all the way through. For instance, *Drinking Wine at Nanting Pavilion in Spring*, *Drinking to Commemorate Yuanjiu*, *Drinking Wine*, *Drinking by the Flowers*, *Songs After Drinking*, *Drinking Wine with Vice-minister Han at Zhengjia Garden*, *Notes after Drinking*, *Raving after Drinking*, *Poem after Drinking with Cui Cundu*, *Commemorating Yuanjiu After Drinking Wine with Li Shiyi*, *Lonely Drinking by the Flowers*, *Answering a Toast*, *the Strong Wine*, *Spring Wine* and *Two Poems after Drinking Wine*. In addition, *A Pipa Song*, one of Bai Juyi's most famous long poems, was composed when he was drinking wine while listening to the beautiful melody of Pipa. Bai Juyi was famous not only for his bibulosity, but also for his wine-making skills. When he was serving as an official, he made great efforts to research wine-brewing methods. During the brewing process, he did not just give orders, rather he personally participated in the work.

- 白居易像

白居易，字乐天，唐朝伟大的现实主义诗人，中国文学史上负有盛名且影响深远的诗人和文学家。他的诗歌题材广泛，形式多样，语言平易通俗，有"诗魔"和"诗王"之称。

A Portrait of Bai Juyi

Bai Juyi, courtesy name Letian was a great realistic poet of the Tang Dynasty, as well as a renowned poet and writer. His poetry involved broad themes and many variations, and his language was simple and easy to understand. He was cordially nicknamed the "Poetry Devil" and the "Poetry King".

- 《琵琶行》郭诩（明）

A Pipa (a Chinese Lute) Song, by Guo Xu （Ming Dynasty, 1368-1644）

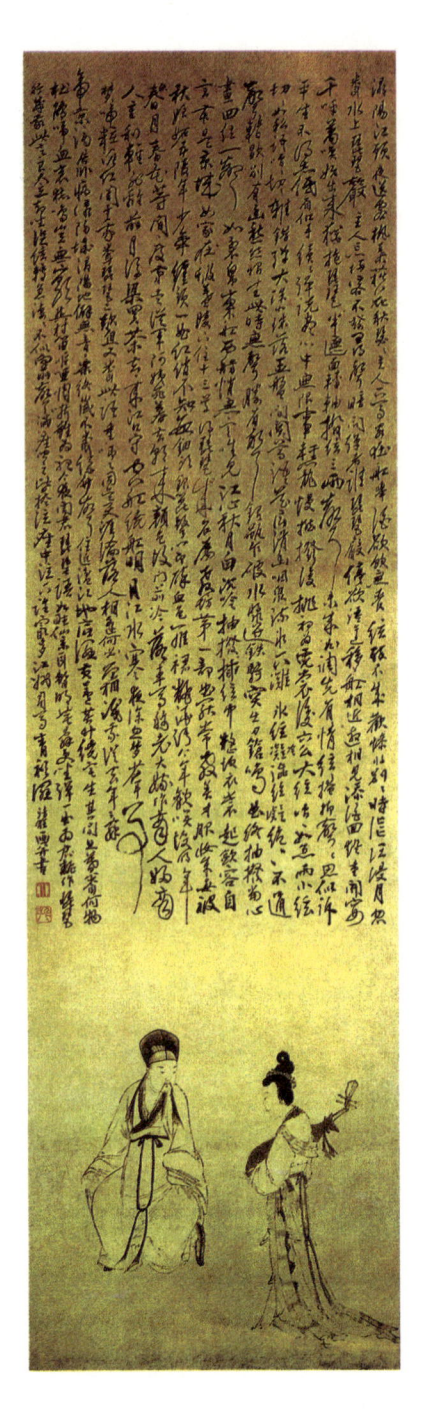

"醉翁"，他的那篇著名的《醉翁亭记》，从头到尾一直贯穿一股酒气。"醉翁之意不在酒，在乎山水之间也。山水之乐，得之心而寓之酒也。"（醉翁的意趣不在于喝酒，而在于欣赏山水的景色。欣赏山水的乐趣，领会在心里，寄托在酒上。）无

People of the Song Dynasty enjoyed a rich and colorful urban life. After attending banquets and drinking wine, scholars composed many excellent poems and verses that fully demonstrated Song people's interest in wine. The renowned verse by Su Shi, a famous scholar of the Northern Song Dynasty, was *Prelude*

水调歌头

苏轼（北宋）

明月几时有，把酒问青天。不知天上宫阙，今夕是何年。我欲乘风归去，又恐琼楼玉宇。高处不胜寒，起舞弄清影，何似在人间！转朱阁，低绮户，照无眠。不应有恨，何事长向别时圆。人有悲欢离合，月有阴晴圆缺，此事古难全。但愿人长久，千里共婵娟。

词人形象地描绘了皓月当空的情景，用一种与明月对话的口吻，在对话中探讨人生的意义。词中表达了词人对超凡脱俗人生境界的向往和对人间的眷恋，是一种复杂而又矛盾的情感，最后抒发了其对生活的热爱和积极向上的人生态度。

Prelude to the Water Melody

Composed by Su Shi (Northern Song Dynasty)

When will the moon be clear and bright?
With a cup of wine in my hand, I ask the blue sky.
I don't know what season it would be in the heaven on this night.
I'd like to ride the wind to fly home.
Yet I fear the crystal and jade mansions are much too high and cold for me.
Dancing with my moon-lit shadow,
It does not seem like the human world.
The moon rounds the red mansion, stoops to silk-pad doors, and shines upon the sleepless.
Bearing no grudge,
Why does the moon tend to be full when people are apart?
People may have sorrow or joy, be near or far apart.

The moon may be dim or bright, wax or wane.
There has been nothing perfect since the golden days.
May we all be blessed with longevity.
Though far apart, we are still able to share the beauty of the moon together.

The author of the lyric vividly described the scene where the bright moon was in the sky. In a dialogue register between the man and the moon, the lyric explored the meaning of life. This lyric expressed the poet's yearning for an extraordinary state and his attachment to the human world. This was a complex and controversial feeling. The last part of the lyric expressed his love for life and a positive attitude toward life.

● 苏轼像

苏轼，号东坡居士，北宋文学家、书画家，唐宋八大家之一。

A Portrait of Su Shi

Su Shi, known as Dongpo Jushi, was a writer, calligrapher and painter of the Northern Song Dynasty. He was also one of eight famous masters of the Tang and Song dynasties.

酒不成文，无酒不成乐。天乐地乐，山乐水乐，皆因有酒。

诗人陆游有一首《钗头凤》，抒写了其与表妹唐琬之间缠绵悱恻的情感，令古今多少人唏嘘不已。词起首一句提到的就是黄縢酒，这是一种用黄纸封盖酒坛的酒，酒启封后色泽动人。

to the *Water Melody*, which provided people with a romantic and cool feeling.

The great scholar Ouyang Xiu was also a well-known heavy drinker. Both his poems and his proses emitted a strong aroma of wine, especially the famous prose *The Pavilion of an Old Drunkard*. The famous lines read: "The drunkard's heart was not in the cups,

• 欧阳修像

欧阳修，字永叔，号醉翁，又号六一居士，北宋卓越的文学家、史学家。

A Portrait of Ouyang Xiu

Ouyang Xiu, courtesy name Yongshu, known as the Drunk Man, also known as Liuyi Jushi, was an outstanding writer and historian of the Northern Song Dynasty.

but in the mountains and waters. The joy of appreciating the landscape was perceived by the heart and carried in his wine." Without wine, there would be no writings and no joy. The joy of heaven, earth, mountains and rivers should all be attributed to wine.

The marathon love story between the famous poet Lu You and his cousin Tang Wan touched the hearts of many generations of the Chinese people because of a verse composed by Lu You titled *The Phoenix Hairpin*. Yellow-Rattan Wine was mentioned in the first line of the poem. This wine is stored in sealed jars covered with rice paper. When opened, the wine disclosed its beautiful luster.

陆游与唐琬

　　陆游是南宋时期的大词人,他与唐琬从小就青梅竹马,情投意合,感情深笃。不料婚后第二年唐琬就被婆婆逐出家门。陆游又被迫娶妻,而唐琬也改嫁他人。十年后,两人在绍兴城外的沈园中不期而遇。两人虽然重逢,却又无法彼此面诉离情。陆游伤心之余,就在园子的壁上题下了一首哀怨的《钗头凤》:

　　红酥手,黄縢酒。满城春色宫墙柳。东风恶,欢情薄。一怀愁绪,几年离索。错,错,错!

　　春如旧,人空瘦。泪痕红浥鲛绡透。桃花落,闲池阁。山盟虽在,锦书难托。莫,莫,莫!

　　后来,唐琬见了这首《钗头凤》后,感慨万千,亦提笔和了一首:

　　世情薄,人情恶,雨送黄昏花易落。晓风干,泪痕残。欲笺心事,独倚斜阑。难,难,难!

　　人成各,今非昨,病魂常似秋千索。角声寒,夜阑珊。怕人寻问,咽泪装欢。瞒,瞒,瞒!

　　不久,唐琬就因忧伤过度而死。

- 陆游像
A Portrait of Lu You

Lu You and Tang Wan

The Southern Song-dynasty *Ci*-poem writer Lu You married Tang Wan. The young couple loved each other dearly. But Lu You's mother could not stand her daughter-in-law and forced her son to divorce her. It was a typical tragedy resulting from an arranged marriage. Several years later, Lu You accidentally ran into his ex-wife in Shenyuan Garden, a private garden located in Shaoxing City, Zhejiang Province. Tang Wan by then had remarried. The accidental reunion did not give them the chance to express their feelings towards each other. Lu You was so sad that he wrote a poem *The Phoenix Hairpin* on the wall of the garden to express his pains about the divorce:

You offered me a cup of yellow rattan wine,
With your pink creamy hands.
The whole town was flooded with spring,
Palace walls were covered with willows green.

- 沈园内墙壁上的《钗头凤》
Verses Inscribed on the Wall of the Shenyuan Garden *The Phoenix Hairpin*

Alas, the harsh east wind broke up our love!
Full of deep sorrow,
I have tasted the separation,
For long years after our parting.
A great error, a great error,
It's a great error beyond repair!
The beauty of spring remains,
Yet you are thin for love in vain.
Your red handkerchief is soaked through,
With tears stained with rouge.
Now peach blossoms have fallen;
Deserted here are the ponds and pavilions.
Even though our love pledges remain true,
There's no way to pass our love letters through.
Everything's over, everything's over,
My everlasting regret is beyond cure!

Later on when Tang Wan saw this poem, she was so touched that she wrote another poem to the same tune in response:

The ways of the world are as thin as a veil,
The ways of men seem to be vile.
Dusk goes away in the rain,
Flowers are easy to fall.
At dawn the wind blows high,
Tear stains have run dry.
I wish to voice my sorrow in a mail,
Leaning on the fence, I chat alone,
Hard, hard, hard!
I can't voice my sorrow at all!
We go each our ways,
Gone are our past days.
My sick soul haunts me as a swing's rope.
The sad horn sounds chilly,
In this dead silence of night.
For fear my sorrow may be found,
I try to put up a show of joy,
But I have swallowed my tears.
Hide, hide, hide!
How can I voice this sad tale at all!

Soon after this, Tang Wan died of over-depression.

> 酒与古典文学四大名著

《三国演义》《水浒传》《西游记》《红楼梦》是中国古典文学中的四大名著。在这四部著名的长篇小说中，人物与故事的发展，以及情节的安排中，都少不了酒这个重要的角色。

《三国演义》

《三国演义》由元末明初小说家罗贯中所著，是中国第一部长篇章回体历史演义小说。这部小说描写了三国时期吴、魏、蜀三个政治集团之间的政治和军事斗争。酒在"三国"中，成为政治家、军事家手中的得力武器，全书中写"以酒谋事"的"酒计"共28次。

忠义之酒："桃源把酒三结

> Wine and the Four Famous Chinese Novels

Romance of the Three Kingdoms, the Water Margin, Journey to the West, and *A Dream of Red Mansions* are four great classics of the Chinese literature where wine played an indispensable role in the development of the story and the plots.

Romance of the Three Kingdoms

Romance of the Three Kingdoms, written by Luo Guanzhong of the late Yuan Dynasty and early Ming Dynasty, is the first long Chinese historical novel with each chapter headed by a couplet giving the gist of the content. This novel focuses on the political and military battles between the three political power blocs that emerged from the remnants of the Han Dynasty. Wine served as a powerful

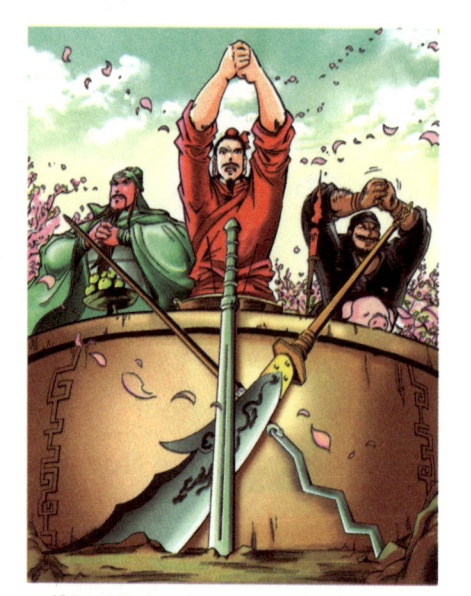

桃园三结义
Oath in the Peach Orchard

weapon for those politicians and militarists in the "three kingdoms". This novel has twenty-eight "wine schemes".

"Oath in the Peach Orchard" is a story about the wine of loyalty. At the beginning of this novel, Liu Bei, Guan Yu and Zhang Fei swore allegiance to the Han Dynasty in the Peach Orchard and pledged to do their best for the country. After the worshipping ceremony, they started drinking until they were totally inebriated. Since then, the three sworn brothers were united as one, and presented a most laudable sentiment of loyalty in ancient China.

"Debating about heroes while heating the wine" is a story about the soul-stirring wine. When Liu Bei stayed in Xuchang City, he attended vegetables in his garden every day in order to avoid Cao Cao's murder. One day, he was suddenly invited by Cao Cao to drink wine. In the middle of the meal, it suddenly started raining. So they started to discuss the topic of heroes using clouds and rain as the metaphor. Pointing at Liu Bei and himself, Cao Cao said: "In current China, there are only two real heroes, who are you and me." Liu Bei was so surprised to hear this that he dropped his chopsticks on the oor. Just

义"。小说开头，刘备、关羽、张飞三人在桃园宣誓结义。祭罢天地，在园里痛饮一醉。这一醉后，三人从此生死同心，演绎出天地间最令人称赞的忠义豪情。

动魄之酒："青梅煮酒论英雄"。刘备在许昌时，为防曹操谋害，每日在花园种菜。他一日忽被曹操叫去喝酒。二人饮酒至半酣，骤雨降至，遂借天上云雨，纵论天下英雄。曹先手指刘后自指，说："今天下英雄，唯使君与操耳！"

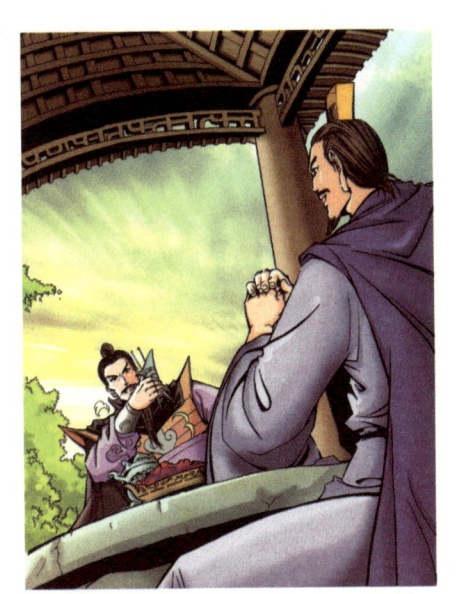

● 煮酒论英雄
Debating about Heroes while Heating the Wine

at that moment, a loud thunder hit. So Liu Bei calmly stooped down and picked up his chopsticks saying: "What a loud thunder! I was scared."

"Guan Yu killed Hua Xiong in the middle of a drinking session" is a story about the heroic wine. When Yuan Shao's army fought Dong Zhuo's army, several officers of Yuan's army were killed by Hua Xiong. At this critical moment, Guan Yu came forward and said: "Keep the wine for me. I will be back in a minute." A few minutes later, he returned with the head of Hua Xiong in his hands. His wine was still warm. What a hero that he was!

"How Zhang Fei defeated Zhang He by pretending to be drunk" is a story about the wise wine. When Zhang Fei's army was fighting Zhang He's army of the Wei Bloc, Zhang He was entrenched in the city for defense. After several dozens of days of confrontation, Zhang Fei started to drink wine every day and when he was drunk he would curse Zhang He in front of the enemy troops. This led Zhang He to believe that Zhang Fei was an alcohol addict who did not understand the military situation. So he launched a surprise attack at night, only to find that he was ambushed.

刘闻言，手中匙箸惊落于地。恰巧这时一声惊雷，刘从容俯首拾箸说："一震之威，乃至于此。"

豪迈之酒："关羽温酒斩华雄"。时值袁绍征讨董卓，被华雄连斩数将。关羽自告奋勇道："酒且斟下，某去就来。"少顷，便提华雄首级入帐。其酒尚温，杯酒间取人首级，如囊中探物。

智慧之酒："张飞佯装酒醉战张郃"。张飞与魏将张郃对战时，张郃固守不出。僵持数十日后，张

109
中国酒文化
Chinese Wine Culture

飞每日饮酒，醉后坐在山前辱骂，给张颌造成张飞好酒不顾军情的假象。于是张颌乘夜偷营，结果中计，受到伏击。

最有名的酒："杜康"。小说中，酒出现得虽然很多，但是酒的牌子却极少出现。最有名的莫过于曹操《短歌行》中的"何以解忧，唯有杜康"。当时曹丞相挟天子以令诸侯，掌控天下的财富，所喝的杜康酒是当时最好的酒了。

《水浒传》

《水浒传》是元末明初编著的长篇小说，是中国第一部用白话文所写的章回体小说。这部小说取材于北宋末年农民起义的故事。作者施耐庵不但喜欢酒、熟悉酒，还善于饮酒，所以他的《水浒传》里处处是酒。

《水浒传》中所提到的酒名有"透瓶香""茅柴白""玉壶春""蓝桥风月""头脑酒"等等。好汉武松过景阳冈喝的是"透瓶香"，又叫"出门倒"。他在大醉之中仅用三拳就把一只老虎给打死了，为民除了害。

The most famous wine in the book is the "*Dukang* Wine". Although many kinds of wine appeared in this novel, very few wine names were mentioned. The most famous wine was the wine in Cao Cao's poem *A Short Song*, reading as "There is nothing but the *Dukang* Wine that can kill my sorrow." At the time, Secretary-general Cao Cao controlled the emperor and commanded the nobles and senior officials. Since he had absolute control over the wealth of the state, Dukang wine that he drank was certainly the best at the time.

The Water Margin

The Water Margin written during the late Yuan Dynasty and the early Ming Dynasty, is the first traditional Chinese novel written in vernacular language, with each chapter headed by a couplet giving the gist of the content. The novel is based on the story of farmers' rebellion during the late Northern Song Dynasty. Shi Nai'an, the writer, was not only fond of wine, familiar with wine, but also good at drinking wine. So his novel was permeated with wine.

The Water Margin mentioned many wine names, such as "Through-Bottle Aroma", "Maochaibai", "Jade Pot

Spring", "Blue Bridge Breeze and Moon" and "Brain Wine". What the famous hero Wu Song drank at the Jingyang Ridge was the wine "Through-Bottle Aroma" which was also called "*Chumendao*", meaning the drinker would fall on the ground as soon as he stepped out of the restaurant. After drinking the strong wine, Wu Song beat a tiger to death in three strikes, eliminating the evil for the local people.

In *The Water Margin*, drinking wine was an expression of courage and bravery. Many heroic figures were interested in drinking wine and eating meat. This increased their determination and boosted their pride. Before Monk Lu Zhishen fought in Wutai Mountain for the first time, he drank ten cups of wine plus "a barrel of wine". After drinking several cups of wine, Song Jiang dared to write an anti-government poem at the Xunyang Restaurant before launching the rebellion. It was by no means an exaggeration to say that many heroes

- 景阳冈武松打虎
Wu Song Beats the Tiger at Jingyang Ridge

- 《水浒传》版画《大闹五台山》
Engraving of *the Water Margin* Story *Riot at the Wutai Mountain*

《水浒传》中人物喝酒表现出的是一种勇猛和气概。众多英雄人物无酒不欢，大碗喝酒，大块吃肉。花和尚鲁智深第一次大闹五台山，吃了十来碗酒后，"又吃了一桶酒"。连谨小慎微的宋江，在造反前几杯酒下肚，竟也敢在浔阳楼上题写反诗。可以毫不夸张地说，《水浒传》中的众英雄有好多是因酒而汇集在一起的。

《西游记》

《西游记》由明代小说家吴承恩所著，描写的是孙悟空、猪八戒、沙和尚护送唐僧西天取经，途中历经九九八十一难的传奇历险故事。虽是写几位僧人西天取经，却也饮酒，当然，饮的是素酒。

小说中，唐僧严守戒律，堪称典范。但在陷空山无底洞被困时，三藏听悟空的话，在"危急存亡之秋，万分出于无奈"，为了哄住妖精，尽管是素酒，也只能"没奈何吃了"。

孙悟空为骗取罗刹女（即铁扇公主）的宝扇，变作牛魔王来到翠云山芭蕉洞。罗刹女为夫整酒接

in this novel converged at Liangshan Mountain because of wine.

Journey to the West

Journey to the West was written by Wu Cheng'en, a famous novelist of the Ming Dynasty. The novel depicted the legendary adventure of a Tang Priest Sanzang with his three disciples — the Monkey, the Pig and the Friar Sand. The four-member team had to overcome eighty-one difficulties and obstacles on the way to the west where the Buddhist Sutra was available. Although they were Buddhist disciples they drank wine on the trip. Of course what they drank was vegetarian wine.

In this novel, the Tang priest was a very pious monk who observed Buddhist rules earnestly. But, when he was imprisoned in the Bottomless Cave of the Empty Mountain, he took the Monkey's suggestion and "had some vegetarian wine" just to fool the devil.

In order to cheat Lady Luocha (Princess Iron Fan) and obtain her magic fan, the Monkey transformed into the image of Lady Luocha's husband the Bull Demon King and came to the Banana Cave of the Blue Cloud Mountain, where Princess Iron Fan warmly entertained him

• 《西游记》版画
画中唐僧念起紧箍咒，头戴紧箍的孙悟空痛得在地上打滚。
Engaving of the *Journey to the West* Story
Tang Priest reads out his magic incantation to tighten the ring on the head of the Monkey to torture him. The Monkey is so painful that he rolled on the ground.

with delicious food and wine. The fake Bull Demon King "had to take the wine cup" in order to fulfill his mission.

A Dream of the Red Mansions

A Dream of the Red Mansions was a type of traditional Chinese novel with each chapter headed by a couplet giving the gist of the content written by Cao Xueqin of the Qing Dynasty. Against the backdrop of the vicissitudes of four big families, Jia, Shi, Wang and Xue, the novel depicted the love and marriage tragedy between Jia Baoyu, Lin Daiyu and Xue Baochai, as well as trifles in the Grand View Garden. The novel demonstrated the irreversible trend of the demise of the feudal society. Author Cao Xueqin was famous for his bibulosity. He adopted the pen name "Mengruan — dreaming about Ruan", expressing his extraordinary admiration for Ruan Ji, a

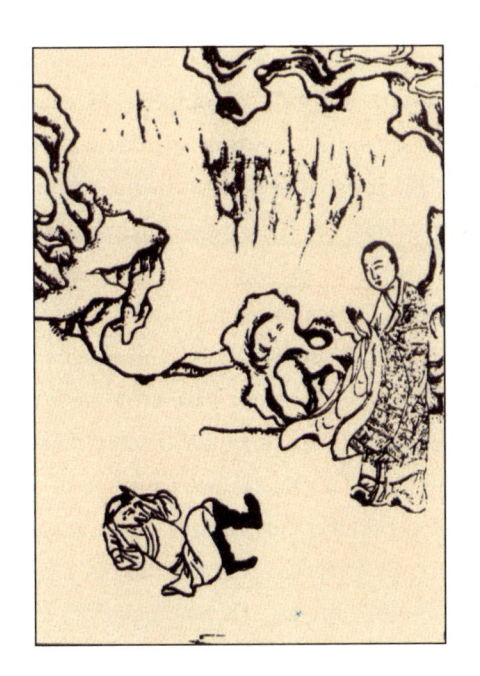

风，擎杯奉上。假牛魔王"不敢不接"，先是"不敢破戒，只吃几个果子，与他言言语语"，实不能推辞，为达到目的只好相陪。

《红楼梦》

《红楼梦》是由清代小说家曹雪芹撰写的章回体小说，全书以贾、史、王、薛四大家族的兴衰为背景，描写了贾宝玉与林黛玉、薛宝钗的爱情婚姻悲剧及大观园中的点滴琐事，揭示了穷途末路的封建社会必将走向灭亡的趋势。作者曹雪芹好饮是出了名的，自号"梦阮"，表达了对魏晋名士阮籍的仰慕。阮籍是有名的"酒徒"，嗜酒如命，听说步兵营藏着三百石美酒，便求为步兵校尉，人称"阮步兵"。他曾痛饮六十天，醉得不省人事，借此躲过政治上的麻烦。而曹雪芹好饮不亚于阮籍。朋友诗中咏及他的生活，有"卖画钱来付酒家""举家食粥酒常赊"等句，可见酒在曹雪芹生活中是不可或缺的。

celebrity of the Wei and Jin dynasties. Ruan Ji was well-known for his love for wine. One day, when he heard there were three hundred jars of good wine kept in a military camp, he immediately begged to serve as the captain of that camp. So he was nick-named the "Infantry Ruan". Once he drank continuously for sixty days until he totally passed out. He did so to avoid some political troubles. Cao Xueqin was by no means a lesser drinker than Ruan Ji. Some of his friends composed poems to describe Cao Xueqin's life as: "He sold paintings to pay for his wine"; "He only had porridge for his family, but he bought wine on credit"; etc. It is evident that wine was indispensable in Cao Xueqin's life.

A Dream of the Red Mansions also depicted the "elegance" of wine drinking. In this book, wine parties were mentioned in about seventy places. Whenever there were drinking parties, there were poems and wine games. As a way of

- 曹雪芹雕像
  A Sculpture of Cao Xueqin

《红楼梦》表现了饮酒的"雅"。写酒宴七十多处，凡饮酒，必作诗，必行酒令。酒令是中国人在筵宴上助兴取乐的饮酒游戏，诞生于西周，完备于隋唐。该书第四十回写到鸳鸯作令官，喝酒行令的情景，展现了当时上层社会喝酒行雅令的风貌。

entertainment during wine parties, wine game first appeared during the Western Zhou Dynasty, and matured during the Sui and Tang dynasties. Chapter forty of the book described the wine game scene whereby Yuanyang was nominated as the commander of the table and the rest played the wine game while drinking. What was shown was how the elite class played the elegant wine game.

- 《红楼梦》中的酒宴场景
  A Wine Feast Scene in the novel *A Dream of the Red Mansions*

金鸳鸯三宣牙牌令

Servant Girl Yuanyang Announcing Game Orders

京剧《贵妃醉酒》

酒是中国戏曲中不可缺少的重要因素，有许多戏以酒或醉酒构成全剧的主要情节。为了表现醉酒的形象和神态，写意性的中国戏曲创造了许多生动逼真的表演程式，营造别开生面的艺术氛围，增强观众的艺术感受。

京剧《贵妃醉酒》本以表现杨贵妃酒醉之后自赏怀春的心态为主，20世纪50年代，梅兰芳去芜存精，改编剧本，使其成为京剧中的经典之作。剧中，杨贵妃从掩袖而饮到随意而饮，梅兰芳以外形动作的变化来表现这位失宠贵妃从内心苦闷、强自作态到不能自制、沉醉失态的心理变化过程。

Beijing Opera *The Drunken Beauty*

Wine is an important element in Chinese operas. Wine or wine-drinking constitute the key plots in many operas. In order to depict the image and demeanor of a drunkard, Chinese operas have invented many vivid performing patterns, and created an artistic atmosphere through spectacular means, both of which have enriched the artistic experience of the audience.

The original version of the Beijing Opera piece *The Drunken Beauty* was mainly about Concubine Yang's self-appreciation and longing for love after she got drunk. In the 1950s, Mei Lanfang adapted the script, converting into a masterpiece of the Beijing Opera repertoire. In the revised version, through the changing body movements, Mei Lanfang (the actor) depicted the psychological changing process of the disgraced concubine from inner anguish to pretentious calm to the explosion of her true emotions.

• 梅兰芳

梅兰芳，京剧大师。他出身于梨园世家，在长期的舞台实践中，对京剧旦角的唱腔、念白、音乐、舞蹈、服装、化妆都有所创造发展，形成了自己的艺术风格，世称"梅派"。经典代表作品有《贵妃醉酒》《霸王别姬》等。

Mei Lanfang

Mei Lanfang, was a Beijing Opera Master. Born in a performing artist family, Mei Lanfang was able to form his own artistic style in the rendition of singing, recitation, music, dancing, and costume, cosmetics, based on his long years of practice. His style was known as the "Mei School". Mei Lanfang's classic representative works include *The Drunk Beauty* and *Farewell to the Beloved Concubine*.

醉拳、醉剑、醉棍

在中国武术中，有与酒结合的武术套路——醉拳、醉剑、醉棍。这些武术系模拟醉态编定，表演者并不是真醉，而是模拟醉者的形态，在"醉"中攻防，似乎醉得站都站不稳，然而在跌撞翻滚之间，随势进招，使人防不胜防。有名的醉拳套路有"醉八仙""太白醉酒""武松醉酒"等。醉拳、醉剑、醉棍不仅具有特殊的攻防性，更具备独具特色的观赏性，是武术中重要的表演项目。

Drunken Boxing, Sword Dance and Cudgel Play

Chinese Martial arts entail wine-related skills such as drunken boxing, drunken sword dance and drunken cudgel play. These performances are choreographed imitating the movements of drunkards. The performers are not really drunk, but they can vividly imitate the gestures of drunkards, and attack and defend when they are "drunk". The performers look like they cannot even stand on their feet, yet they can suddenly attack the opponent while rolling and stumbling. The famous program routines include "Drunken Immortals", "Drunken Poet Li Bai", "Drunken Hero Wu Song", etc. With a special power for attack and defense, drunken boxing, drunken sword dance and drunken cudgel play also have their unique visual attraction. Indeed they are important members of the Chinese Martial Arts family.

- 中国功夫——醉剑 (图片提供：FOTOE)
Chinese Martial Art—Drunken Sword Dance

> 酒与古代书画

中国古代的书法家、画家大都嗜酒，这些艺术家们往往多愁善感，追求浪漫的生活，具有鲜明的个性，故多借酒兴进行创作。酒有兴奋中枢神经的作用，可以使人精神亢奋，才思敏捷，能够激发出创作的灵感。

东晋"书圣"王羲之醉时挥毫而作《兰亭序》，"遒媚劲健，绝代所无"，而至酒醒时"更书数十本，终不能及之"。

唐代僧人怀素嗜酒，至酒兴起，在寺院的墙壁屏障、衣服器具之上写字，人谓之"醉僧"。他酒醉泼墨，方留下神鬼皆惊的《自叙帖》。李白写怀素："吾师醉后依胡床，须臾扫尽数千张。飘飞骤雨

> Wine and Ancient Calligraphy and Painting

Most of the ancient Chinese calligraphers and painters were wine lovers. They tend to be sensitive and emotional, pursuing romantic lives and distinct personalities. Wine could stimulate the central nervous system and make them excited, enlightened and inspired.

Wang Xizhi, the *Calligraphy Master* of the Eastern Jin Dynasty, wrote the *Preface to the Poems Composed at the Orchid Pavilion* when he was drunk. It was observed by some that "the calligraphy was so beautiful and powerful, unique and unprecedented". When he sobered up, "he wrote many more copies. Unfortunately, none was as good as the original one".

Tang Dynasty Monk Huaisu was

《自叙帖》【局部】怀素（唐）
Autobiography Copybook for Calligraphy [part] by Huaisu (Tang Dynasty, 618-907)

addicted to wine. After drinking, he often wrote Chinese characters and words on the temple walls, screens, clothes and utensils. So people called him the "Drunkard Monk". It was wine that helped him complete the amazing masterpiece *Autobiography Copybook for Calligraphy*. Li Bai once wrote a poem about Monk Huaisu, saying that "My master was drunk lying beside the bed. In no time he wrote thousands of calligraphy pieces. The ink dropped on paper like heavy rainfall and the paper fell on the ground like snowflakes."

Tang Dynasty calligrapher Zhang Xu was famous for his "highly Cursive Scripts". After over-drinking, his inspiration would be triggered. He got so excited that he could not help shouting and running in his house. He would take a brush and start writing calligraphy, or dip hair in the ink and write with his hair. When he sobered up, he was also surprised to see the work he had done while drunk. Thanks to wine, he was able to produce the famous masterpiece of *Four Ancient Poems*.

Wu Daozi, a famous Painting master of the Tang Dynasty, reputably known as "Wu's Blowing Strips", would not start the painting work until he was

惊飒飒，落花飞雪何茫茫。"

以"狂草"传于后世的唐代书法家张旭，大醉后，灵感骤至，状态极佳，异常兴奋，在庭内不住地狂呼疾走，或抓笔在手一挥而就，或以头发浸墨而书。醒后，看到所书之效果，连自己也觉得神异，于是有"挥毫落纸如云烟"的《古诗四帖》。

"吴带当风"的唐代"画圣"吴道子作画前必酣饮大醉，醉后为画，挥毫立就。"元四家"中的黄公望更是"酒不醉，不能画"，醉笔染丹青。

明代大画家郑板桥的字画千金

deeply inebriated. Then he could finish a piece of painting in no time. Another painter Huang Gongwang, one of the four most accomplished artists of the Yuan Dynasty, was also a heavy drinker. He would not be able to paint until he is drunk. Indeed, "no wine, no work".

It was difficult to get a painting or

"吴带当风"

"吴带当风"是对吴道子人物画风格的概述。吴道子所作的人物画突出人体曲线和自然的结合，人物的衣袖和飘带具有迎风起舞的动势，故有"吴带当风"之称。后人以此赞美其高超的画技与飘逸的画风。

"Wu's Blowing Strips"

The expression "Wu's Blowing Strips" vividly summarized the style of Wu Daozi's painting of human figures. A prominent feature of his paintings was the harmonization of the lines of the human body with that of nature. Sleeves and strips of the figures were so dynamic as if they were blowing in the wind. Later generations used the expression "Wu's Blowing Strips" to praise his excellent skills and the style of his paintings.

- 《孔子像》吴道子（唐）
  A Portrait of Confucius, by Wu Daozi (Tang Dynasty, 618-907)

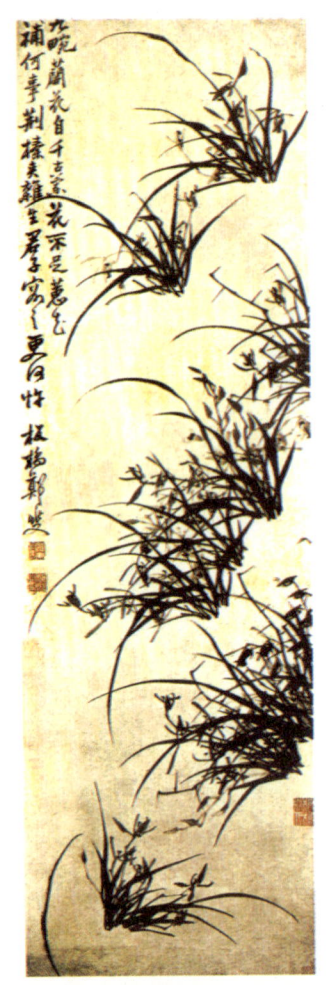

• 《兰草图》郑板桥（清）
The Orchids, by Zheng Banqiao (Qing Dynasty, 1616-1911)

calligraphy by Zheng Banqiao, a famous painter of the Ming Dynasty. So the requestors would bring wine to treat Zheng Banqiao before asking him for the favor. Zheng Banqiao usually satisfied them when he was drunk. Although Zheng was fully aware of these people's tricks, he could not resist the temptation of the delicious wine. So he wrote a self-mocking poem about it: "I prefer to appreciate the bright moon alone, but I cannot wait for the wine to come. I laugh at those who come to request paintings from me. They have to wait until I am totally drunk."

Tang Yin (courtesy name Bohu), was a Chinese painter and literati of the Ming Dynasty. He was famous for his poems, prose, painting and calligraphy, with painting as his best. He claimed to be the "number one talent in southern China". When he was thirty years old, he was accused of connections with a bribery case in the national civil service examination. So he was put in jail and his civil service career was spoiled forever. Since then he pursued a life of pleasure and earned a living by selling his paintings. Later, he built a hut named the "Peach Blossom Cottage" at the Peach Blossom Dock where he also planted

难求，于是求画者拿美酒款待，趁其醉意，求其字画。郑板桥也知道求画者的把戏，但耐不住美酒的诱惑，只好写诗自嘲："看月不妨人去尽，对月只恨酒来迟。笑他缣素求书辈，又要先生烂醉时。"

　　明代书画家、文学家唐寅，字伯虎，诗文书画都有名，尤以画著称，人称"江南第一才子"。他三十岁时受科场案牵连入狱，功名受挫，从此致力绘事，以卖画为生。后来，他在桃花坞建了茅舍，名为"桃花庵"，还种了几株桃树，并为此赋诗。

some peach trees and composed a poem for it.

- **《蕉林酌酒图》陈洪绶（明）**
  此画以石桌上厚厚的书卷和巨大的酒瓮为依托，来烘衬主人持杯细酌的闲情雅致，将主人所用的各种器物刻画得极为细致。如旁边木雕上所放之双耳铜鼎、石桌上带鸬鹚杓的大酒坛、主人手中之爵杯和侍女手托之酒壶等，皆精美非常。尤其是主人手中之爵杯，显然是一件仿古爵杯。

  Drinking in the Banana Grove, by Chen Hongshou (Ming Dynasty, 1368-1644)
  The painting uses stacks of books and a huge wine jar as the background to highlight the protagonist character's elegant leisure reflected in his contemplated drinking and savoring of the wine. All the wine utensils are meticulously depicted. The bronze tripot with two ears, the big jar on the stone table with a bird-shaped scoop, the Jue cup in the hands of the hero, and the wine flagon in the hands of the servant girl, all were exquisite. The wine cup in the hands of the protagonist character is obviously an imitation of an ancient *Jue* cup.

A Dream under the Tung Tree, by Tang Bohu (Ming Dynasty, 1368-1644)

桃花庵歌

桃花坞里桃花庵，桃花庵下桃花仙；
桃花仙人种桃树，又摘桃花卖酒钱。
酒醒只在花前坐，酒醉换来花下眠；
半醉半醒日复日，花开花落年复年。
但愿老死花酒间，不愿鞠躬车马前；
车尘马足富者趣，酒盏花枝贫者缘。
若将富贵比贫贱，一在平地一在天；
若将贫贱比车马，他得驱驰我得闲。
别人笑我忑疯癫，我笑别人看不穿；
不见五陵豪杰墓，无花无酒锄做田。

The Song of Peach Blossom Cottage

In the Peach Blossom Land there is a peach blossom Cottage;
A peach blossom lover lives in Peach Blossom Cottage.
The peach blossom lover plants peach trees in days fine;
He sells his peach blossoms for money to buy wine.
When he is not drunk, he would sit before the flowers;
He would lie beneath them to spend his drunken hours.
From day to day half-drunk, half-sober he'd appear;
The peach flowers blossom and fall from year to year.
I would grow old and die among flowers and wine;
Rather than bow before the steed and carriage fine.
The rich may love their dust-raising carriage and bowers;
The poor only enjoy their cups of wine and flowers.
If you compare the poor with the rich low and high,
You will find the one on the earth, the other in the sky.
If you compare the poor with the carriage and steed,
The poor have leisure while the rich gallop with speed.
Others may pit me so foolish and so mad;
I laugh at them for those who can't see through are sad.
Can you find where the tombs of gallant heroes stand?
Without flowers or wine, they are into plough land.

- 《韩熙载夜宴图》顾闳中（五代 南唐）

《韩熙载夜宴图》描绘的是山谷五代时南唐大官僚韩熙载骄奢淫逸夜生活的一个场面。韩熙载的父亲韩光嗣被后唐李嗣源所杀，韩熙载被迫投奔南唐，官至史馆修撰兼太常博士。韩熙载雄才大略，屡陈良策，希望统一中国，但频遭冷遇，使其对南唐政权失去信心。不久，北宋雄兵压境，南唐后主李煜任用韩熙载为军相，妄图挽回败局，韩熙载自知无回天之力，却又不敢违抗君命，于是采取消极抵抗的方式，沉溺于酒色。李煜得知韩熙载的情况，派画院待诏顾闳中、周文矩等人潜入韩府，他们目识心记，绘成多幅《韩熙载夜宴图》。现保存在北京故宫博物院的这幅《韩熙载夜宴图》展现了古代豪门贵族"多好声色，专为夜宴"的生活情景。图中的注子、注碗的形制是研究酒具发展变化的重要资料。

Han Xizai's Night Feast by Gu Hongzhong (Southern Tang of the Five Dynasties, 937-975)

The painting depicts the extravagant night life of a senior official Han Xizai of the Southern Tang of the Five dynasties. Han Xizai whose father was killed by Li Siyuan of the Later Tang Dynasty, had no choice but to join the Southern Tang, where he was assigned a senior official post in charge of the renovation of the archives, updating of historical records, as well as education. As a very talented officer with a great vision for the unification of the middle kingdom China, Han made many good strategic suggestions to the King, who unfortunately ignored these suggestions. Eventually, Han lost confidence in the Southern Song Dynasty. Not long after, the Northern Song Dynasty assembled its military forces at the border area. In an attempt to rescue the situation, Li Yu of the Southern Tang appointed Han Xizai as the military prime minister. Knowing that it was no longer possible to reverse the situation and at the same time daring not to disobey the King's order, Han Xizai adopted a strategy of passive resistance by indulging in wine and women. Having learnt about Han Xizai's behavior, King Li Yu sent imperial painters Gu Hongzhong and Zhou Wenju to Han's house to peek at what Han was doing. Based on their observation and recollection, these artists sketched many paintings depicting *Han Xizai's night feasts*. The painting which is kept in the Palace Museum in Beijing depicts the night life of nobilities in ancient times. The wine-warming ewer and the bowl depicted in the paintings are good reference material for the research of the evolution of wine utensils.

> 酒与酒具

中国古代有很多造型优雅的酒具，这是中国酒文化的重要组成部分。古代酒具的材质主要有陶瓷、青铜、玉石、金银、竹木牙角等几类。

陶瓷酒具

陶质酒具是中国人最早使用的酒具。在距今大约4500年的新石器时代晚期的大汶口文化遗址中出土了具有饮酒功能的彩陶，但当时的陶器并不一定只有单一的功能，一般的食具如碗、钵等也有可能是酒具。而且远古时期的酿酒工艺没有过滤这一步骤，所酿的酒醪呈糊状，并不适合口较小的器皿。

随着酿酒业的发展，酒具从一般食具中分化出来成为可能。在

> Wine and Wine Utensils

There were many elegant and exquisite wine utensils in ancient times, comprising an important part of the wine culture in China. Raw materials of these utensils included mainly ceramics, bronze, jade, gold, silver, bamboo, wood, ivory, horn, etc.

Wine Ceramic Utensils

Pottery utensils were the oldest wine utensils. Color potteries with the function of drinking wine were excavated from the Dawenkou Cultural Site of the late Neolithic Age (4500 years ago). However, it was possible that other pottery utensils for food such as bowls or cups were also used to contain wine. A good argument to support this view was that since wind brewing procedures

新石器时代晚期，各地就已出现许多用途明确的酒具。在种类繁多的陶杯中，有平底杯、圈足杯、高柄杯、觚形杯等。陶质酒具到殷商时期就逐渐被青铜酒具所取代。

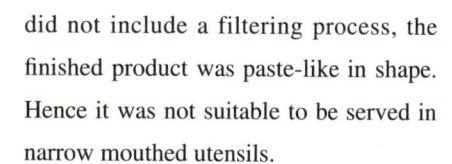

did not include a filtering process, the finished product was paste-like in shape. Hence it was not suitable to be served in narrow mouthed utensils.

It was only the development of the wine industry that made it possible to separate wine utensils from other food utensils. During the late Neolithic Age, many specific wine utensils were made including pottery cups, flat-bottom and round bottom, long-handled cups and goblet-shaped cups. Pottery wine utensils

- 高柄杯（大汶口文化）

A Long-handled Cup（Dawenkou Culture, 4500 B.C.-2500 B.C）

- 单耳杯（大汶口文化）

A Single-ear Cup（Dawenkou Culture, 4500 B.C.-2500 B.C）

- 黑陶宽把杯（良渚文化）

A Wide-handled Black Pottery Cup（Liangzhu Culture, 5600-4900 B.C.）

瓷器从陶器发展而来，大致出现于东汉前后，一直沿用至今。瓷器与陶器相比，其胎质更致密，外观更漂亮，而且便于清洗，经久耐用，因此瓷质酒具的性能远超陶质酒具。

受酿酒工艺的影响，早期酿造的是度数较低的黄酒，在饮用时需要加热，所以单次饮用量较大，酒具也较大。到唐代时，度数较高的白酒走进了人们的生活，使单次饮用量减少，酒具的形制也比过去小了许多。另外，唐代出现了桌子，也出现了一些适于在桌上使用的酒

were gradually replaced by bronze utensils during the Yin-Shang Dynasty.

Emerged around the time of the Eastern Han Dynasty, porcelain utensils were developed from pottery and have been used until modern times. Compared to pottery utensils, porcelain utensils were far finer in texture, more exquisite in design, easy to clean and more enduring. In short, porcelain utensils were far superior to pottery utensils.

Due to limitations of the wine brewing technology, only *Huangjiu* of low alcoholic content was produced in ancient times. Because *Huangjiu* had to be heated when served, larger utensils were needed as people usually took large quantities. It was not until the Tang Dynasty that high alcohol content *Baijiu* entered people's life. People began to drink less with smaller cups or containers. In addition, during the Tang Dynasty, with the emergence of tables, wine utensils suitable to be placed on the table appeared as well, eg. ewer (the shape is similar to a

- 酒桌（明）
酒桌的出现也带动了酒具的发展。
A Wine-drinking Table (Ming Dynasty, 1368-1644)
The emergence of tables promoted the development of wine utensils.

具，如注子（形状似今日之酒壶，有喙，有柄，既能盛酒，又可注酒于酒杯中）。宋代是瓷器生产的鼎盛时期，出现了注子和注碗的配套酒具组合。明代的瓷制酒具以青花、斗彩较有特色，清代瓷制酒具以珐琅彩、素三彩、青花玲珑瓷及各种仿古瓷为主。

current day wine agon with a beak and a handle. It can be used to contain wine or to serve wine in cups). The Song Dynasty witnessed a remarkable development of porcelain production. Wine sets composed of ewer and bowl became available. While porcelain wine sets during the Ming Dynasty were featured by blue and white porcelain and polychrome porcelains, the porcelain wine utensils during the Qing Dynasty included mainly color enamels, plain tri-color, blue-white exquisite porcelains and various antique imitation porcelains.

- 龙泉窑注子（明）
A Wine-warming Ewer Made at the Longquan Kiln (Ming dynasty, 1368-1644)

- 白瓷注子（唐）
A White Porcelain Wine-warming Ewer (Tang Dynasty, 618-907)

青铜酒具

　　青铜器出现于新石器时代晚期，其中青铜酒具延续使用了一千六百余年。尤其是在商周时期，青铜铸造技术大大提高，青铜酒具的品种数量之多、纹饰之精美、工艺水平之高超，都达到了惊人的程度。

　　在夏代和商代早期，青铜器就逐渐被用于生产和生活中。在夏代的二里沟文化遗址中，出土了爵、角、斝、觚、盉等青铜酒具，形制复杂，器物表面已出现饕餮纹、乳丁纹、云雷纹等花纹。商代早期青铜酒具的总体风格与夏代相似，但

• 象尊（商）
An Elephant-shaped Wine Container *Zun*
(Shang Dynasty, 1600B.C. – 1046B.C.)

Wine Bronze Utensils

Bronze utensils appeared during the late Neolithic Age. Bronze wine utensils had been used for 1,600 years, especially during the Shang and Zhou dynasties when bronze casting technology was greatly improved. The large quantity, exquisite design and decoration and excellent processing technology all reached an amazing level of sophistication.

　　During the Xia and early Shang Dynasty, bronze tools and utensils gradually emerged in production activities and people's daily life. Bronze wine utensils including tri-legged cups, horn-shaped cups, cups with ears, goblets and *He* were excavated from Erligou Culture Site of the Xia Dynasty. These wine utensils had complex patterns such as taotie, dots and clouds on the surface. Although bronze wine utensils made during the early Shang Dynasty were very similar to those made during the late Xia Dynasty in terms of the overall style, they were more variations and sophisticated in production technology. The walls of the utensils became thicker, the patterns (especially taotie pattern) more complex, and the decoration method more refined.

豕尊（商）
A Swine-shaped Wine Container *Zun*
(Shang Dynasty, 1600B.C. – 1046B.C.)

种类明显增多，制作工艺也更为先进，器体胎壁逐渐变厚，纹饰（特别是饕餮纹）也开始变得繁复，装饰手法更为精细。

商代中期到西周早期是青铜文化的鼎盛期，这一时期青铜工艺有了长足的发展。从出土实物来看，商代中晚期的酒具器形硕大厚重，纹饰多为复层花纹（两层以上的花纹），且变化多样，常见的有兽面纹、云雷纹、鳞纹、龙纹、连珠纹、夔龙纹、鸟纹、鱼纹、蛇纹、虎纹、芭蕉纹等，富丽堂皇，彰显尊贵。这一时期还有各种铸造精美的动物形酒器，如牛尊、羊尊、豕尊、象尊、犀尊、鸮尊、鸮卣、豕卣等。西周早期的酒具延续了商代晚期的风格，不同的是，西周酒具上有铭文者增多，铭文的文字量也增多，多的可达十几个或几十个，甚至上百个。

Bronze ware production enjoyed a period of rapid development during the mid-Shang Dynasty to early Western Zhou Dynasty. From unearthed objects, it is evident that bronze utensils produced during the mid and late Shang Dynasty were huge and heavy with complex patterns (two or more layers), such as beast-head, cloud, squama, loong, pearl, bird, fish, snake, tiger and banana. They look magnificent and elegant. This period also witnessed the production of many exquisite, animal-shaped wine containers, such as shapes of ox, sheep, pig, elephant, rhinoceros and owl. Wine utensils produced during the Western Zhou Dynasty followed the style of the late Shang Dynasty in general. The difference was that they had more inscriptions and characters on the utensils. Some had as many as several dozen or even one hundred inscribed characters.

Entering the mid and late Western Zhou Dynasty, the production of bronze

到了西周中晚期，青铜酒具的铸造工艺逐渐衰退，少有精品。纹饰开始趋于朴素简单，而且也越来越粗糙，铭文也出现脱漏、颠倒的情况。春秋战国时期，战争频繁，礼崩乐坏，加上铁器的出现，使得青铜酒具不再受到重视。在制作工艺方面，出现了一些新的因素，如错金错银、镶嵌宝石等装饰手法；纹饰以蟠螭纹为主，还出现了写实纹饰。到了秦汉时期，酒具材质以金银器、釉陶、瓷器等为主，青铜酒具就逐渐退出了历史舞台。

wine utensils suffered serious setbacks as a result of the declining casting technology. Only a few excellent pieces were produced during that period. Patterns became simplified and rough, and words were wrong or missing in the inscriptions. Due to frequent battles during the Spring and the Autumn and the Warring States periods, the collapse of the etiquette system and the emergence of iron utensils, bronze wine utensils lost their prominent position. In terms of production technology, there were also some innovations, such as the decoration method of mixing gold, silver and inlaid gems; a legendary animal- panchi pattern and realistic patterns appeared on these utensils. During the Qin and Han dynasties, wine utensils were mostly made of gold, silver, glazed pottery and porcelains. Bronze wine utensils gradually disappeared from the drinking tables.

- 妇好鸮尊（商）
Queen Fuhao's Bird-shaped Wine Container *Zun* (Shang Dynasty, 1600B.C. – 1046B.C.)

常见的青铜酒具器形
Commonly Seen Bronze Wine Utensils

尊

尊在古代曾是各类酒器的泛称，作为一种具体器物指的是大型高体的盛酒器。尊主要有大口尊和觚形尊两种。大口尊为大口，折肩，圈足，有圆形和方形两种；觚形尊为侈口，筒形，形制与觚相似，但较觚更为高大。

Zun (Wine Container)

Once upon a time, *Zun* was the generic name for various kinds of wine containers in ancient China. As a specific name, it refers to the huge and tall wine container. *Zun* can be divided into two major sub-categories: one is those with a large-mouth, and the other has the shape of a goblet. The large-mouthed ones are round or rectangular in shape, with folded shoulders and circular legs and can be further divided into two types of round and square shaped; the goblet-shaped ones have an exaggerated month and column-like body, like that of a goblet, only bigger.

方彝

方彝是一种大型盛酒器，形状像房子，器身方形，高身如墙壁、盖似屋顶，造型华丽典雅。

Fangyi

A *Fangyi* is a huge wine container that looks like a house. It's square in shape with its sides like walls and its lid like the roof. It's both magnificent and elegant.

- 兽面纹方尊（商）
A Beast-face Pattern Square Wine Container *Zun* (Shang Dynasty, 1600B.C.-1046B.C.)

- 牛首兽面纹尊（商）
A Baffalo-head Shaped with Beast Face Pattern Wine Container *Zun* (Shang Dynasty, 1600B.C.-1046B.C.)

- 父戊方彝（商）
Fuwu Fangyi (Shang Dynasty, 1600B.C.-1046B.C.)

壶

壶是一种盛酒器，基本形制为有盖、长颈、有耳、圆腹、圈足。式样颇多，有椭圆形壶、圆腹壶、圆角长方形壶、方形腹壶、扁鼓形壶等。

Flagon

A Flagon is a wine container with basic structures including a lid, long neck, ear, round belly and circular legs. There are many different shaped flagons such as oval, round-bellied, rectangular with round corners, square-bellied and flat-bellied.

卣

卣是一种盛酒器，多为圆口或椭圆形口，深腹，圈足，有盖。卣与壶形体相似，但整体较壶矮，且口径较大。绝大多数卣有提梁。

You

A *You* is a wine container usually with a round or oval mouth, deep belly, circular legs, and a lid. While *You* is similar to Flagon, it is only shorter, with a bigger mouth and a handle.

瓮、瓿

瓮、瓿是一类盛酒器，基本形制是大口，宽肩，圆腹，高圈足，与罍、尊相似，但较罍矮，较尊口小。

Weng and Bu

Weng and *Bu* are a kind of wine container with the basic structure of a big mouth, wide shoulders, round belly and high circular legs. Similar to *Lei* or *Zun*, *Weng* and *Bu* are shorter than *Lei*, and their mouths are smaller than those of the *Zun*.

- 革带纹扁壶（战国）

A Flat Flagon with Belt Patterns (Warring States Period, 475 B.C.-221 B.C.)

- 祖辛铜卣（商）

King Zuxin's Bronze *You* (Shang Dynasty, 1600B.C.-1046B.C.)

- 四羊首瓿（商）

A Four Sheep-head *Weng* (Shang Dynasty, 1600B.C.-1046B.C.)

罍

罍是大型贮酒器，其基本形制是敞口，宽肩，肩上有两耳，口径小于肩径，正面腹下有环鼻，可系绳提起，便于倒酒，平底或圈足，个别有类似屋顶的盖。

Lei

A *Lei* is a huge container for storing wine. Its basic structure includes an open mouth, wide shoulders and two ears. Underneath the front belly, there are nose rings that allow the utensil to be lifted with a string. It has a flat bottom or a ring foot. Some have roof-shaped lids.

- 羊首兽面纹铜罍（商）

A *Lei* with Sheep-head Patterns (Shang Dynasty, 1600B.C.-1046B.C.)

觥

觥为盛酒或饮酒器，其基本形制有椭圆形和方腹形，腹部表面纹饰繁缛。腹上有盖，多做成有角的兽头或长鼻上卷的象形。器身有流，流较短且上扬，有的盖与流共同组成一个兽头。

Gong

A *Gong* can be used as a wine container or a drinking cup whose basic structure includes an oval or square belly, with complicated patterns on it. There is usually a lid on the belly top made into the shape of a beast's head with horns or an elephant with a rolling long nose. There is also a duct on the body, usually short and upright, sometimes, the duct and the lid form a beast's head together.

- 作册折觥（西周）

A Folding *Gong* (Western Zhou Dynasty, 1046B.C.-771B.C.)

盉

盉是温酒与调酒器，其基本形制是深腹，敛口，前有管状流，后有盖，三足或四足。

He

A *He* is used for wine-warming or wine-mixing. Its basic structure includes a deep belly, a narrow mouth, a front duct, a lid and three or four legs.

- 吴王夫差盉（春秋）

King of State Wu Fuchai's *He* (Spring and Autumn Period, 770B.C.-476B.C.)

爵

爵是温酒、饮酒器，纹饰丰繁秀美。其基本形制是口部前有流，后有尾，中部为深圆腹杯，下有三锥足，左侧有供提的鋬，流与杯口之间有两小柱。

Jue

A *Jue* can be used to heat up the wine or to drink from it. It usually has complicated and beautiful patterns. Its basic structure includes a duct in front and a tail behind and a deep belly in the middle. It has three pointed legs and a handle on the left. There are also two small poles between the duct and the mouth.

角

角（音"绝"）是温酒、饮酒器，形似爵，不同的是角无柱无流且两翼若尾，尺寸普遍大于爵，有的带盖。

Jue

Another kind of *Jue* is also used for heating up above-mentioned wine or drinking from it. It is very similar to the *Jue* mentioned above, only it does not have the two small poles and the duct, and the two wings look like tails. It is usually bigger than the above-mentioned *Jue* and some have lids.

斝

斝为煮酒或温酒器，形似爵却比爵大，无流无尾，却有两柱，口圆，扁平鋬或圆鋬；底有平底、圜底、鬲形底之分；以三棱足居多，亦有款足，足内空者为多。

Jia

A *Jia* is used to cook or to heat wine. Its shape is similar to the smaller Jue, only it is bigger. It has no duct, no tail, but two poles and a flat or round handle. The bottoms are flat, circular or caldron shape. Most of these utensils have three hollow legs.

- 父乙爵（西周）
A Buffalo-head-shaped *Jue* (Western Zhou Dynasty, 1046B.C.-771B.C.)

- 父乙铜角（西周）
A Buffalo-head-shaped *Jue* (Western Zhou Dynasty, 1046B.C.-771B.C.)

- 方斝（商）
A Square *Jia* (Shang Dynasty, 1600 B.C.-1046B.C.)

觚

觚是饮酒器，其基本形制为侈口、束腰、长身，口和足部似喇叭口状。

Gu

A *Gu* is used for drinking wine. Its basic structure includes an outwardly extending mouth, narrow waist and long body. Both the mouth and the legs look like a bugle.

觯

觯是饮酒杯，其形似侈口的小壶，样式有圆体与扁体两种。

Zhi

A *Zhi* is a drinking cup which looks like a small-sized flagon. There are round *Zhi* and flat *Zhi*.

枓、勺

枓、勺是斟酒器。枓有柄，古人认为其像北斗七星之形，曾写作"斗"，为与量器的斗相区别，而用"枓"字；勺在造型上与枓接近，皆有小杯与柄，但二者也存在一定的区别。一般枓为曲柄，勺为直柄。

Dou and Shao

Both the *Dou* and the *Shao* are dippers for serving wine. Both have handles and a small scoop on one end. The only difference is that *Dou*'s handle is bending while *Shao*'s handle is straight.

- 兽面纹觚（商）
A Beast-face Pattern *Gu* (Shang Dynasty, 1600B.C.-1046B.C.)

- 庶觯（西周）
A *Zhe Zhi* (Western Zhou Dynasty, 1046B.C.-771B.C.)

- 蟠虺纹枓（春秋）
A Wine Serving *Dou* with Snake Patterns (Spring and Autumn Period, 770B.C.-476B.C.)

玉石酒具

早在新石器时代，中国的远古先民就已经选用玉材制器，但因人们对玉的尊崇，长期以来，玉制品都不是为实用而制，几乎都是作为礼器与佩饰使用。大约到了秦汉时期，才开始出现实用玉器，而其中大部分具有实用价值的玉器也仍然是作为礼器或陈设使用，如高足玉杯、耳杯、角形杯等。

到了唐代，产于甘肃酒泉市的夜光杯因王翰的一首《凉州词》而备受追捧。早在西周时期，夜光杯已是西域贡品。当时夜光杯是用和田玉制成，后因运输路途遥远，故改用酒泉玉来制作夜光杯。

Wine Jade Utensils

As early as during the Neolithic Age, ancient people began to make and use jade utensils. But because of people's respect and worship of jade, jade utensils were mostly used for rituals or accessories rather than practical use. It was not until the Qin and Han dynasties that jade utensils for practical use appeared. Still most of these utensils were for rituals or for display purposes, for instance, the long-leg jade cups, ear cups, horn shaped cups, etc.

Entering the Tang Dynasty, the luminous jade cup produced in Jiuquan City, Gansu Province earned its fame largely because of a poem written by Wan Han entitled *Song of Liangzhou Frontier*. As early as in the Western Zhou Dynasty, luminous cups were offered by western region rulers as tribute items to the emperor. Initially the cups were made of jade produced in Hetian. Later, because of the long distance of transportation, people used jade produced in Jiuquan to make luminous cups.

• 白玉羽觞（唐）

耳杯，又名羽觞，呈雀鸟状，平底呈椭圆形，因其左右形如两翼故名。饮酒时，以双手执耳杯饮酒。

A Wine White Jade Cup (Tang Dynasty, 618-907)

A bird-shaped eared cup with an oval-shaped flat bottom. It got the name because of the two wing-shaped ears. The drinker has to hold the two side ears when drinking.

- **渎山大玉海（元）**

又名"黑玉酒瓮"，是用整块杂色独山玉琢成的贮酒瓮，周长约5米，重3500公斤，可贮酒30石。据传是元世祖忽必烈在至元二年（1265年）运来置于在琼华岛上，宴赏功臣。现存北京北海公园前团城玉瓮亭内。

A Grand Wine Jade Container from Dushan Mountain (Yuan Dynasty,1206-1368)

Also named "Wine Black Jade Jar", is carved from a whole piece of Dushan jade. With a circumference of five meters, the container weighs three thousand and five hundred kilograms. It can store thirty *Dan* of wine. As legend has it, this container was shipped by Kublai Khan of the Yuan Dynasty to the Jade Islet in Beihai Park in 1265. He used it to contain wine for banquets when rewarding well performed officials. Now the container is stored in the Jade Jar Pavilion of the Circular City in Beijing's Beihai Park.

凉州词

王翰（唐）

葡萄美酒夜光杯，欲饮琵琶马上催。

醉卧沙场君莫笑，古来征战几人回。

将士们正准备畅饮夜光杯里的葡萄美酒，马背上传来催促出征的琵琶声。请不要笑话我们醉卧沙场，自古以来出征打仗有几个人能回来？

这首诗以豪放的风格描写了战士饮酒作乐的情景，具有浓郁的边塞军营生活色彩。

Song of Liangzhou Frontier

By Wang Han (Tang Dynasty)

The cups of jade would twinkle with wine of grapes at night. Drinking to pipa songs, we are summoned to fight. Don't laugh if we lay drunk on the battle ground! How many warriors ever come back safe and sound?

Warriors are about to drink wine from a luminous cup, when the pipa-lute begins to play, calling warriors to go to the battle field. Do not laugh at us if you see us lying drunk on the battlefield. How many soldiers could return from a battlefield since ancient times?

This poem depicts the scene of soldiers enjoying drinking at a frontier barrack in a bold style.

明清时期的玉石饮酒器种类和数量都有所增多，不仅有玉杯、玉高足杯、玉耳杯、玉壶等功能明确的酒具，还有许多兼有礼器功能的仿古彝器，如玉尊、玉觯、玉角、玉卣、玉觚等。

金银酒具

金银是古代中国最贵重的金属，早在春秋战国时期就被用来制作酒具，供上层贵族使用。从出土器物的情况来看，春秋战国时期的金银酒具为数不多，但在工艺上已经达到较为完善的程度。特别是以金、银合为一体制作而成的酒具，工艺先进，令人惊叹不已。

到了唐代，金银酒具开始兴盛起

The Ming and Qing dynasties witnessed an increase in wine utensils made of jade both in variety and in quantity. There were not only function-specific utensils such as jade cups, jade goblets, jade ear cups and jade flagons, but also imitations of ancient utensils for rituals, such as jade *Zun*, jade *Zhi*, jade *Jue*, jade *You*, jade *Gu*, etc.

Gold and Silver Wine Utensils

Gold and silver were the most precious metals in ancient times in China. As early as during the Spring and Autumn Period, gold and silver were used to make wine utensils for nobles. Analysis of the excavated utensils showed that there were not many gold and silver wine utensils during the Spring and Autumn Period although the production technology was already quite sophisticated. In particular, wine utensils made of gold and silver

- 金樽（战国）
  八棱形长身酒器，器身两侧刻有攀缘相向的立体龙纹，一龙首向下，用白银镶双翼；一龙首向上，用白银镶双角，金银相映，奇趣无比。龙眼以蓝琉璃镶嵌，有画龙点睛之效。

  A Gold *Zun* (Warring States Period, 475 B.C.-221 B.C.)
  This is a gold octagonal prism-shaped long wine container with verticle loong patterns on the sides. One loong heads up while the other one heads down. One loong has silver wings and the other one has silver horns, very eye catching. The eyes of the loongs are inlaid with blue glass, serving as the last touch.

• 错银鎏金壶（战国）
A Gld-gilded Flagon (Warring States Period，475 B.C.-221 B.C.)

来，其品种之繁多、花纹之精美是前所未有的。常见的器形有金杯、银杯、高足金杯、金壶、银壶等，杯体外壁常常铸有精美纹饰，具有富丽堂皇的时代风格。

demonstrated the advanced processing technology of the time and they were really amazing.

Gold and silver wine utensils made during the Song Dynasty were more exquisite in shape and had more vigorous

• 高足金杯（隋）
大口，口沿外翻，高足中空喇叭状，足柄上端先焊有一圆片，再焊合于杯身。
A High-legged Gold Cup (Sui Dynasty, 581-618)
This cup has a large open mouth, like the shape of a trumpet.

• 鎏金刻花摩羯纹银长杯（唐）
A Gilt Silver Drinking Vessel with Carved Design and Makara Pattern (Tang Dynasty, 618-907)

宋代金银酒具在前朝的基础上不断创新，造型更加玲珑，纹饰更具生活气息。此后的元明清三朝，金银酒具又以华丽、繁复的宫廷风格为主，且工艺越来越精细。

designs as they had been improved on the basis of preceding dynasties. In the ensuing Yuan, Ming and Qing dynasties, gold and silver wine utensils adopted the imperial style of magnificence and were made with increasingly refined craftsmanship.

• 带托金酒注（明）
A Gold Wine Pot with Tray (Ming Dynasty, 1368-1644)

• 金座玉爵（明）
A Jade *Jue* on Gold Base (Ming Dynasty, 1368-1644)

- 金錾花云龙纹葫芦执壶（清）
A Gourd-shaped Wine Pot with Flower, Loong and Cloud Patterns (Qing Dynasty, 1616-1911)

- 乾隆款金爵（清）
A Gold *Jue* of Qianlong Period (Qing Dynasty, 1616-1911)

- 光绪款金执壶（清）
A Gold Flagon of Guangxu Period (Qing Dynasty, 1616-1911)

竹木牙角酒具

竹木牙角是天然材质，容易取得，但竹、木因质地容易朽坏而留存至今的不多。竹木酒具流传至今的，多为竹、木胎经髹漆工艺处理的。

Bamboo, Wood, Ivory and Horn Wine Utensils

Bamboo, wood, ivory and horn are easily available natural materials. However, as bamboo and wood easily go rotten, only a few bamboo or wood utensils that had been treated through a painting process

牙角通常是指象牙和犀角，不过用象牙和犀角制作的酒具也很少。较早的象牙酒具有出土于商代墓葬中的象牙筒形杯。汉唐以后，犀、象在中国绝迹，直到宋明时与南亚诸国贸易往来，犀角和象牙进口才多了起来。古人认为犀角能解百毒，故多用来做成酒具饮酒以养生。

明清时期犀角酒具造型多样，以杯盏为多，也有碗、盂、爵等；纹饰内容丰富，有花卉、山水、人物，以及仿古的蟠螭纹等，有的犀角杯内壁也雕有纹饰，颇为华丽。

survived the time.

Ivory and horns usually refer to elephant tusks and rhinoceros horns; however, there are only a few wine utensils made of ivory and horns. Some ivory utensils were excavated from Shang-dynasty tombs. When elephants and rhinoceros disappeared in China during the Han and Tang dynasties, wine utensils made of ivory and horns also disappeared. It was not until the Song and the Ming dynasties when China engaged in commercial transactions with South-Asian countries that ivory and horn products reappeared in China. Since ancient Chinese people believed that rhinoceros horns could detoxify all kinds of poison, they made wine utensils with horns to maintain good health.

The Ming and Qing dynasties witnessed diverse wine utensils made of rhinoceros horns, including cups, calyxes, bowls, *Jue*, etc. There were rich patterns, including flowers, mountains, streams, figures and archaized patterns on the utensils; some even had patterns on the interior wall.

- 犀角螭龙纹杯（清）
Rhinoceros Horn Cup with Loong Pattern (Qing Dynasty, 1616-1911)